Probability Activities

For Problem Solving & Skills Reinforcement

By Robert Lovell

KEY CURRICULUM PRESS
Innovators in Mathematics Education

Copyright © 1993 by Key Curriculum Press. All rights reserved.
Published by Key Curriculum Press, 1150 65th Street, Emeryville, CA 94608.
Graphics by Ann Rothenbuhler. Illustrations by Joe Spooner.

Printed in the United States of America   14 13 12 11 10   05 04 03 02   ISBN 1-55953-067-7

About The Author

Robert Lovell received a degree in Mathematics from Rose-Hulman Institute of Technology in 1963. After a brief career in industry, he began teaching high school mathematics. In 1968 he received a Master's degree from Indiana State University. He has attended numerous National Science Foundation (NSF) institutes.

A classroom teacher since 1965, Mr. Lovell has taught at North Side High School in Fort Wayne, Indiana since 1970. During this time he has been interested in improving the curriculum for the non-college bound.

Mr. Lovell has been a freqent speaker at local, state, and national mathematics conferences. He has been active in the National Council of Teachers of Mathematics (NCTM) and has had articles published in the *Mathematics Teacher*. In 1983 he was the Indiana recipient of the Presidential Award for Excellence in the Teaching of Mathematics. In 1985 his alma mater, Rose-Hulman, granted him the Outstanding High School Teacher Award.

Introduction

● PHILOSOPHY

For years, students in the non-academic high school mathematics curriculum have been subjected to a rehash of middle school mathematics. Middle school texts and high school General Mathematics texts are indistinguishable. Chapters on operations with whole numbers, fractions, and decimals are common to all of these texts. Additionally, chapters on percents, area and perimeter, and probability and statistics generally appear. Many of the exercise sets consist of 25 to 30 of the same type of exercise. These books expect mastery of a topic in one lesson. Most high school students are turned off by the material. "We did this last year!" is a common and valid complaint.

Five years ago, I began writing material that addressed the shortcomings of the typical General Mathematics curriculum. Probability is a topic that students have an intuitive feeling for, yet it can provide "disguised drill" in the area of fractions, decimals, and percents. And students can explore "higher" mathematics (Pascal's Triangle and expected value) even if they have limited backgrounds. Using this book, you need never expect mastery in one day. Once a topic is introduced, it appears again in the problem sets in that unit and occasionally in future units.

Each student must have a simple calculator. However, the problem sets clearly address the issue of when the calculator is an appropriate tool to use.

I've tried to provide students the opportunity to develop specific problem-solving techniques. Also, problems from The Guinness Book of Records® appear in about half of the sixty lessons. These are motivational, provide for some interesting discussions, and are true problems. Motivational material in the form of palindromes, magic squares, alpha-numerics, brain busters, and calculator capers (the answer is a word spelled upside down) appear in many lessons.

Most lessons are intended to be covered in one class period. However, it will take most classes a semester to cover all of the material.

PROCEDURES

My colleagues and I have used this material since 1987. During that time, we've developed procedures that have been successful with our students. So that you don't need to reinvent the wheel, I offer the following suggestions:

STUDENT MATERIALS

Each student needs a three-prong folder with pockets. They use the paper in it to take notes and to work through examples I place on the board or overhead. I grade it at the end of each unit. After it is graded, I instruct students to clean it out and start over for the next unit. Many students have trouble organizing this folder. To assist them, I find it helpful to keep a running list on the board or on a 22 x 28 piece of tagboard of what should be in the folder. For example,

UNIT ONE

1. 9/5   Notes: Approximating percents
                                       Egore and His Coins

2. 9/6   Notes: Probability definition
3. Quiz 1
4. 9/8   Experiment 1

On the day before I grade the folder, I provide students with a checklist of what should be in the folder. The checklist is identical to the list on the board but with two columns where the student and I each check whether the material is there.

HOMEWORK

I assign homework almost daily. It must be graded on a regular basis. There are various procedures you can use: exchange and grade papers in class, have an aid grade them, or grade them yourself. Spot checking papers is sometimes sufficient.

TESTS AND QUIZZES

Quizzes are usually provided after every two lessons. They are meant to be scored on a ten point basis. I find administering quizzes at the beginning of the period effective. Students who are late receive a "0." Students who are absent usually have time in class on a later date to make them up. I require students to keep the graded quizzes in their folders. A 100 point test is provided for each unit.

GRADING

Here's a suggested weighting:

Tests	40%
Quizzes	20%
Homework	20%
Folder	20%

TABLE OF CONTENTS

Lesson	Description
27	• Mentally determining 5% and 15% of a number • Differentiating between events that lead to trees with equal and unequal length branches
28	• Drawing tree diagrams for two stage events in which the two trials are dependent
29	• Drawing tree diagrams for three stage events in which the three trials are dependent
30	• Review **Unit 3 Test**
31	• Determining probabilities in two stage events without drawing trees
32	• Solving problems by *Looking for a Pattern*
33	• Writing decimals as percents rounded to the nearest tenth
34	• Number tricks
35	• Developing Pascal's Triangle and applying it in probability situations
36	Experiment 6: Tossing Six Coins
37	• Determining the percent of a number with a calculator
38	• Using a calculator to solve a word problem involving percents
39	• Non-probability problems leading to Pascal's Triangle
40	• More non-probability problems leading to Pascal's Triangle **Unit 4 Test**
41	Experiment 7: Calculating Average Loss in a Game • Determining the expected value of a game
42	• Detemining the value of a game when all outcomes have positive payoffs • Determining the elements in a three by three magic square
43	• Determining the elements in a four by four magic square
44	• Mentally determining percent increase/decrease
45	Experiment 8: Chuck-a-Luck
46	• Determining the value of a game using Pascal's Triangle
47	• Determining expected value when the game involves a tree with unequal length branches • Determining percent increase/decrease with acalculator
48	Experiment 9: Roulette
49	• Determining the mean of a set of numbers
50	• Review **Unit 5 Test**
51	• Determining the median of a set of numbers
52	• Constructing a frequency distribution • Calculating the mean and median from a frequency distribution
53	• Determining the range of a set of data
54	Experiment 10: Drawing Cards to Get an Ace
55	• Creating palindromes

Lesson	Description
56	• Constructing ordered stem and leaf plots
	• Determining upper and lower quartiles
57	• Constructing back to back stem and leaf plots
58	• Constructing box and whisker plots
59	• Drawing scatterplots
	• Determining trend lines and correlations
60	• Review
61	• Review
	Unit 6 Test

Description of Units

Unit 1

PROBABILITY

Simple events

1. Definitions of probability both intuitive and mathematical
2. Probability of simple events using dartboards, spinners, decks of cards, and tables
3. Meaning of *and* and *or*
4. Probabilities of complementary events

PROBLEM SOLVING

Make a List

EXPERIMENT

Spin a Spinner

PERCENT

1. Converting fractions whose denominators are factors or multiples of 100 to percents
2. Approximating percents by writing as fractions with denominators close to 100
3. Converting percents to fractions

This unit consists of ten lessons, four quizzes, and a unit test.

Unit 2

PROBABILITY

Multi-stage independent events where the outcomes in each stage are equiprobable

1. Listing outcomes of two-stage events (rolling two dice, spinning a spinner twice, etc.) in a table and determining probabilities from the table
2. Drawing tree diagrams of multi-stage events in order to list all outcomes and to determine the probabilities

PROBLEM SOLVING

Guess and Test

EXPERIMENTS

1. Rolling a Pair of Dice
2. Tossing Three Coins

PERCENT

1. Writing decimals as percents
2. Using the calculator to convert fractions to percents
3. Solving word problems of the following type:

   Of 56 students interviewed, 43 preferred Cheeryos. What percent perferred Cheeryos?

This unit consists of nine lessons, four quizzes, and a unit test.

Unit 3

PROBABILITY

Multi-stage independent events where the outcomes in each stage are not equiprobable and multi-stage dependent

1. Drawing tree diagrams, listing probabilities on the branches and multiplying the probabilities to determine the probability of a particular outcome
   a. trees with equal length branches
   b. trees with unequal length branches
   c. independendent events
   d. dependent events
2. Determine the expected number of times an outcome should occur in n trials where n is a "nice" number

PROBLEM SOLVING

Logical Reasoning with Alpha-Numerics

EXPERIMENTS

1. Drawing Two Colored Cubes from a Cup (replacement).
2. Using a Random Number Table to Simulate a Three Game Series

PERCENT

1. Mentally calculating 0%, 5%, 10%, 15%, 25%, and 50% of a number
2. Writing percents as decimals
3. Using a percent line to estimate answers to percent problems

This unit consists of eleven lessons, four quizzes, and a unit test.

Unit 4

PROBABILITY

Multi-stage events and binomial probability distributions

1. Determining probabilities without drawing trees
2. Constructing Pascal's Triangle and applying it in appropriate probability situations

PROBLEM SOLVING

Recognizing Patterns

EXPERIMENTS

Tossing Five Coins

PERCENTS

1. Write decimals as percents rounded to the nearest tenth
2. Use a calculator to determine "$p\%$ of b"

This unit consists of ten lessons, three quizzes, and a unit test.

Unit 5

PROBABILITY
      The expected value of a game
1. Determining the expected (average) win/loss
2. Determining the average (mean) of a set of numbers

EXPERIMENTS
1. Is This a Fair Game?
2. Chuck-a-Luck
3. Roulette

PERCENT
      Determining percent increase and decrease.

ENRICHMENT
      Magic Squares

This unit consists of ten lessons, five quizzes, and a unit test.

Unit 6

STATISTICS
1. Determining the mean, median, range, and upper and lower quartiles of a set of data
2. Comparing two sets of data by construction of back to back stem and leaf plots or box and whisker plots
3. Constructing frequency distributions
4. Determining statistical measures from frequency distributions and box and whisker plots
5. Constructing scatter plots to determine correlation between two sets of data

ENRICHMENT
      Palindromes

This unit consists of eleven lessons, four quizzes, and a unit test.

PROBABILITY
Lesson Plan 1

QUIZ

None

OBJECTIVES

1. Introduce the concept that the probability of an event occurring is a number between 0 and 1 inclusive and can be expressed either as a fraction or a percent.
2. Students will define probability.
3. Students will convert fractions with nice denominators to fractions whose denominators are 100, and then to percents.
4. Students will write percents as simplified fractions.

MATERIALS NEEDED

1. Transparencies 1A-C
2. Problem Set 1

CLASS ACTIVITIES

Have each student number from 1–12 on a piece of paper. Show Transparency 1A and have the students estimate probabilities. Answers to some problems should be the same (problem 3) but will vary greatly in others (problem 5). These examples lead to an intuitive definition of probability (Transparency 1B). The other transparency should be used to explain converting fractions to percents and vice versa. The fractions in these examples have denominators that can easily be transformed to 100.

ASSIGNMENT

Problem Set 1

PROBABILITY 1A

Assign a probability to each event

1. You will miss a day of school this year.

2. You will get an A in math.

3. A tossed coin will land "tails."

4. You will gain 10 pounds this year.

5. You will live to be 100 years old.

6. The sun will rise tomorrow.

7. The next President of the United States will be over 21 years old.

8. A card drawn from a standard deck will be a club.

9. You will take a space trip to the moon.

10. The next baby born in this state will be a girl.

11. You will listen to a CD today.

12. You have type "D" blood.

PROBABILITY 1B

Definition

The probability of an event is a number between 0 (0%) and 1 (100%) *inclusive* that indicates the likelihood of the event occuring.

0%	50%	100%
impossible		certain

PROBABILITY 1C

Convert to percents

1. 73/100

2. 14/50

3. 3/10

4. 11/20

5. 41/25

6. 146/200

Convert to simplified fractions

7. 47%

8. 65%

9. 140%

10. 100%

11. 4%

12. 75%

PROBABILITY
Problem Set 1

Write as fractions having a denominator of 100. Then write as a percent.

1. 17/50
2. 14/25
3. 1/2
4. 3/25
5. 3/4
6. 17/20
7. 9/20
8. 2/25
9. 73/50
10. 46/200
11. 66/300
12. 432/400

Write as fractions having a denominator of 100. Then simplify.

13. 60%
14. 25%
15. 100%
16. 45%
17. 200%
18. 40%
19. 150%
20. 48%
21. 22%
22. 10%
23. 160%
24. 75%

Assign a probability (%) to each event.

25. You will graduate from high school.
26. A card drawn from a standard deck is red.
27. You will live to be 20 years old.
28. Next baby born in your city will be a girl.
29. There is intelligent life on Mars.
30. You will miss more than 10 days of school this semester.

GUINNESS RECORD®: The fastest speed attained by a woman on water skis is 111.11 mph by Donna Patterson Brice at Long Beach, California on August 21, 1977.

31. At that rate, how far would she travel in 30 minutes?
32. At that rate, how long (hours and minutes) would it take her to travel a distance of 611 miles?

BRAIN BUSTER: Use the numbers from 1–6 so that all sides of the triangle have the same sum.

PROBABILITY
Lesson Plan 2

QUIZ

None

OBJECTIVES

1. Students will approximate the percent represented by fractions that can be written with denominators close to 100.
2. Students will use the *Make a List* problem-solving technique.

MATERIALS NEEDED

1. Transparencies 2A and 2B
2. Problem Set 2

CLASS ACTIVITIES

Work through Transparency 2A with the class. The students should realize that 57/98 is close to (but slightly greater than) 57%.

Explanation: If a student got 57 out of 98 questions correct on a test, how many would he expect to get correct if there were 100 questions? All of the denominators can be bumped close to 100. Example 5 can be estimated (24/96), but can be determined exactly by first simplifying to 1/4.

Transparency 2B introduces the *Make a List* procedure. It is important that the students use a systematic approach in order to avoid repetitions and omissions.

ASSIGNMENT

Problem Set 2

PROBABILITY 2A

Approximate as percents

1. 57/98

2. 3/11

3. 5/33

4. 17/52

5. 6/24

6. 19/21

7. 68/198

8. 11/9

9. 54/899

10. 3/7

PROBABILITY 2B

Egore and His Coins

Egore has 3 coins in his pocket. They are pennies, nickels, and dimes. List the different amounts of money he might have.

Pennies	Nickels	Dimes	Total

PROBABILITY
Problem Set 2

Write as fractions having a denominator of 100. Then write as a percent.

1. 3/50   2. 1/2   3. 13/25   4. 1/5   5. 17/20   6. 3/4
7. 11/10   8. 18/200   9. 2460/1000

Write as fractions having denominators near 100. Then write as a percent.

10. 13/49   11. 5/9   12. 11/24   13. 11/26   14. 7/33   15. 21/19
16. 5/6   17. 38/202 18. 15/306

Write as fractions having a denominator of 100. Then simplify.

19. 35%   20. 64%   21. 120%   22. 100%   23. 25%   24. 8%

Ivan went to the store and bought a pencil for 23¢. He paid the exact amount and received no change.

25. Make a list of the nine different ways that Ivan could have paid for the pencil.
    Your columns should be:

Dimes	Nickels	Pennies
2	0	3

26. What is the fewest number of coins Ivan could have used?
27. What is the greatest number of coins Ivan could have used?
28. If Ivan used six coins to pay for the pencil, what were they?
29. Could Ivan have paid for the pencil with exactly 8 coins?

Alice threw 3 darts at the dart board. All of the darts hit the board.

30. List the ten possible ways the darts could have landed and the score for that turn.
    Your columns should be:

Five	Ten	Twenty	Total
3	0	0	15

31. What is her highest possible score?
32. What is her lowest possible score?
33. How many different scores could she make?
34. How many ways could she score 30 points?

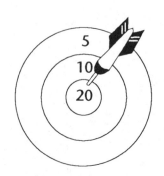

PROBABILITY
Lesson Plan 3

QUIZ

Quiz 1

OBJECTIVES

Students will use the definition of probability of an event to determine the probability of the event occurring.

MATERIALS NEEDED

1. Quiz 1
2. Transparency 3
3. Problem Set 3
4. A set of 10 cards marked as on Transparency 3 (optional)

CLASS ACTIVITIES

Work through the example on the transparency. Emphasize that the denominator is always 10. Answers should be given as fractions and percents.

ASSIGNMENT

Problem Set 3

PROBABILITY QUIZ 1

Name _____

Write each of the following as a percent.

1. 47/100 = _____

2. 3/25 = _____

3. 136/200 = _____

4. 7/5 = _____

5. 17/17 = _____

Write each of the following in simplified form.

6. 100% = _____

7. 48% = _____

8. 350% = _____

Give an estimate of the following probabilities.

9. The next President of the United States will be over 30 years old. _____

10. You will be in school tomorrow. _____

PROBABILITY 3

Definition

The probability of event *A* occurring is given by

$$P(A) = \frac{\text{number of ways } A \text{ can occur}}{\text{total number of outcomes}}$$

Example

A card is drawn.

1	2	3	4	5
6	7	8	9	10

1. P(3)=

2. P(15)=

3. P(odd)=

4. P(not 6)=

5. P(prime)=

6. P(more than 6)=

7. P(striped)=

8. P(less than 20)=

9. P(multiple of 3)=

PROBABILITY
Problem Set 3

Write as fractions with denominators of 100. Then write as percents.

1. 37/50   2. 88/200   3. 11/25   4. 4/5   5. 1/2   6. 19/10

Write as fractions with denominator of 100. Then simplify.

7. 35%   8. 64%   9. 120%   10. 100%   11. 75%   12. 8%

Write as fractions having a denominator near 100. Then estimate the percents.

13. 47/51   14. 15/26   15. 2/11   16. 46/198   17. 47/33   18. 5/3

A card is drawn at random. Write the probabilities as fractions and percents.

19. P(1)   20. P(6)
21. P(plain)   22. P(not plain)
23. P(less than 8)   24. P(odd)
25. P(divisible by 2)   26. P(more than 4)

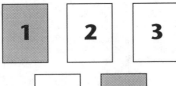

Jim Schue scored nine points in a basketball game. Long shots count three points, field goals count two points, and free throws count one point.

27. Make a list of the 12 ways that Jim could have scored nine points.
28. How many ways could he have scored nine points without scoring a free throw?
29. How many ways could he have scored nine points without scoring a long shot?

BRAIN BUSTER: How many rectangles are in the picture?

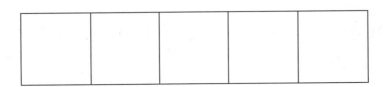

PROBABILITY
Lesson Plan 4

QUIZ   None

OBJECTIVES   Given the probability of an event, students will be able to determine
                        the probability of the complementary event. Probabilities will be
                        given as percents or fractions. Students will be able to interpret the
                        notation P(~A) to mean the probability of not A.

MATERIALS NEEDED   1. Transparency 4
                        2. Problem Set 4
                        3. A set of cards marked as on Transparency 4 (optional)

CLASS ACTIVITIES   Work through the examples on the transparency.

ASSIGNMENT   Problem Set 4

PROBABILITY 4

S	P	I	N	S

A card is drawn at random.

1. P(S) =

2. P(not S) =

3. P(I) =

4. P(~I) =

5. P(Q) =

6. P(~Q) =

7. P(rain) = 30%

   P(no rain) =

8. P(win lottery) = 1/1,000,000

   P(don't win lottery) =

9. P(A) = 3/17

   P(~A) =

10. P(~T) = 21%

   P(T) =

PROBABILITY
Problem Set 4

Write as percents.

1. 3/4   2. 51/50   3. 34/200   4. 183/300   5. 48/300   6. 1/2

Write as fractions having a denominator near 100. Then estimate the percent.

7. 8/49   8. 13/24   9. 156/199   10. 19/33   11. 3/14   12. 2/9

13. The probability of rain is 45%. What is the probability that it will not rain?

14. The probability that Slotorun will win the race is 15%. What is the probability that he will not win the race?

15. The probability of drawing a club is 1/4. What is the probability of not drawing a club?

16. The probability of catching a cold is 3/1000. What is the probability of not catching a cold?

17. The probability of not missing the boat is 23/43. What is the probability of missing the boat?

A card is drawn at random. Write the probabilities as fractions and percents.

18. P(N)
19. P(vowel)
20. P(A)
21. P(comes before Y in the alphabet)
22. P(letter is made of straight lines)
23. P(~N)

N	O	R	T	H

The Aardvarks scored 21 points in a football game. Six points are awarded for a touchdown, three points for a field goal, and one point for an extra point after a touchdown.

24. List all of the different ways that the Aardvarks could have scored 21 points.

25. How many ways could the Aardvarks have scored 21 points without a field goal?

The Zebras scored 30 points in a football game.

26. List all of the ways that they could have scored 30 points.

27. How many ways could the Zebras have scored 30 points without scoring an extra point?

GUINNESS RECORD®: The tallest LEGO tower ever built was constructed in Tel Aviv, Israel in May of 1990. It consisted of 221,560 bricks and was 59.5 feet tall.

28. Approximately how many times taller than you was the tower?

PROBABILITY
Lesson Plan 5

QUIZ   Quiz 2

OBJECTIVES   Students will distinguish between the words *and* and *or* in a
                        probability setting.

MATERIALS NEEDED   1. Quiz 2
                        2. Transparency 5
                        3. Problem Set 5
                        4. A set of cards marked as on Transparency 5 (optional)

CLASS ACTIVITIES   For an *and* statement to be satisfied, both parts of it must be
                        satisfied. For an *or* statement to be satisfied, only one part of it must
                        be satisfied. Transparency 5 gives appropriate examples. The
                        probabilities should be written as fractions and then approximated
                        as percents. The last letter in the first half of the alphabet is *m*.

ASSIGNMENT   Problem Set 5

PROBABILITY QUIZ 2

Name _____

1. Using quarters, dimes, and nickels, list the eight ways you could pay for a 45¢ item.

   Quarters   Dimes   Nickels

2. What is the smallest number of coins that you could use? _____

3. How many ways could you pay for the item without using dimes? _____

Write as percents.

4. 11/25 = _____   5. 222/200 = _____

6. Write as a reduced fraction. 80% = _____

Estimate the following as percents.

7. 35/102 ≈ _____   8. 7/11 ≈ _____

PROBABILITY 5

A card is chosen at random.

M	A	T	H	E	M	A

T	I	C	S

1. P(E) =

2. P(~E) =

3. P(S) =

4. P(vowel) =

5. P(second half of alphabet) =

6. P(second half or vowel) =

7. P(second half and vowel) =

8. P(before I and vowel) =

9. P(before I or vowel) =

PROBABILITY
Problem Set 5

One card is drawn at random. Write the probabilities as fractions and percents.

1. P(8)
2. P(11)
3. P(less than 11)
4. P(dotted)
5. P(even or dotted)
6. P(even and dotted)
7. P(plain and prime)
8. P(plain or prime)
9. P(more than 4 or striped)
10. P(more than 4 and striped)
11. The probability that Stan will marry is 76%. What is the probability that he will not marry?
12. The probability of a dropped piece of bread falling buttered side down is 4/5. What is the probability that it will not fall buttered side down?
13. The probability that the stoplight is red is 3/7, and the probability that it is green is 3/7. What is the probability that it is yellow?

Write as percents.

14. 46/50   15. 3/4   16. 23/20   17. 66/200   18. 66/300   19. 246/200

Write as reduced fractions.

20. 50%   21. 75%   22. 10%   23. 100%

Estimate.

24. 45/49   25. 1/9   26. 4/11   27. 63/299

BRAIN BUSTER: Find the start number.

START   **END**

□ — (x 9) — (+ 6) — (÷ 3) — (− 6) — 5

PROBABILITY
Lesson Plan 6

QUIZ   None

OBJECTIVES   Given a "square dartboard," the student will be able to determine
                        the probability that a randomly thrown dart will land in a specified
                        region of the dartboard.

MATERIALS NEEDED   1. Transparency 6
                        2. Problem Set 6

CLASS ACTIVITIES   In explaining the situation, point out that the darts are thrown
                        randomly but that we'll consider only darts that hit the dartboard.
                        Further explain that we'll assume that the dartboards can be divided
                        evenly into pieces identical to the smallest indicated region. The
                        transparency gives appropriate examples. Separate the first
                        dartboard into fourths and the last two into eighths.

ASSIGNMENT   Problem Set 6

PROBABILITY 6

The Square Dartboard

A dart is thrown at the dartboard. Determine the probabilities.

P(A) =

P(~A) =

P(B or C) =

P(B) =

P(C or D) =

P(D) =

P(~D) =

P(A) =

P(D) =

P(E) =

P(D or C) =

PROBABILITY
Problem Set 6

A dart is thrown randomly at a dartboard. Write each of the probabilities as reduced fractions. (Assume the dart hits the board and that the board can be divided evenly into pieces identical to the smallest indicated region.)

A	B	C
D	E	F

1. P(A)
2. P(~A)
3. P(A or B or C)
4. P(A or E or F)

A	B	C
		D
E	F	

5. P(A)
6. P(B)
7. P(C)
8. P(~C)
9. P(E or F)
10. P(A or E or F)

A	B
C	D

11. P(A)
12. P(C or D)
13. P(D)
14. P(~D)
15. P(C)

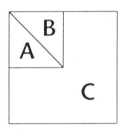

16. P(A or B)
17. P(A)
18. P(~A)
19. P(C)

One card is drawn at random. Write the probabilities as fractions and estimate the percents.

20. P(vowel)   21. P(~vowel)
22. P(E)   23. P(A)
24. P(comes before J in the alphabet)
25. P(comes before J and is a vowel)
26. P(comes before J or is a vowel)
27. P(the letter is made of straight lines)
28. P(T or a vowel)
29. P(T or is made of straight lines)

N O R T H

S I D E

PROBABILITY
Lesson Plan 7

QUIZ   Quiz 3

OBJECTIVES   Given a circular dartboard, the student will be able to determine the
                       probability that a randomly thrown dart will land in a specified
                       region of the dartboard.

MATERIALS NEEDED   1. Quiz 3
                       2. Transparency 7
                       3. Problem Set 7

CLASS ACTIVITIES   The procedures with the circular dartboards are similar to those with
                       the square dartboards. If students have difficulty determining the
                       fractional part of the circle, remind them that we assume the board
                       can be divided evenly into pieces identical to the smallest region.

ASSIGNMENT   Problem Set 7

PROBABILITY QUIZ 3

Name _____

1. Write as a percent. 47/50 = _____

2. Estimate as a percent. 19/26 = _____

3. The probability of Sam missing the bus is 82%. What is the probability that he will not miss the bus? _____

4. The probability that Samantha will win is 13/15. What is the probability that Samantha will not win? _____

| 1 | 2 | 3 | 4 | 5 |

A card is drawn at random. Write each of the probabilities as fractions.

5. P(3) = _____   6. P(~3) = _____

7. P(less than 7) = _____

8. P(even and less than 4) = _____

9. P(even or less than 4) = _____

10. P(a multiple of 3) = _____

PROBABILITY 7

The Circular Dartboard

A dart is thrown at the dartboard. Determine the probabilities.

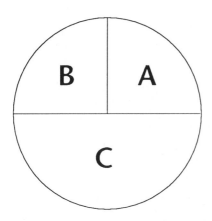

P(A) =

P(~A) =

P(A or C) =

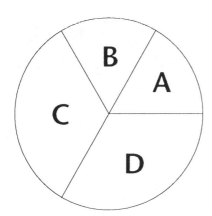

P(A) =

P(~A) =

P(C) =

P(C or D) =

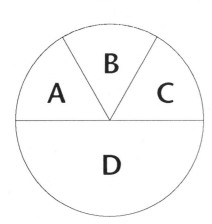

P(B) =

P(A or D) =

P(A or B) =

P(~C) =

PROBABILITY
Problem Set 7

A dart is randomly thrown at a circular dartboard. Write the probabilities as simplified fractions.

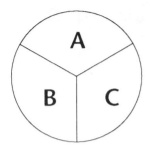 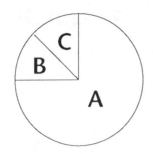 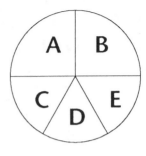

1. P(A)	5. P(A)	9. P(A)
2. P(~A)	6. P(B or C)	10. P(C)
3. P(A or B)	7. P(B)	11. P(~C)
4. P(A or B or C)	8. P(~B)	12. P(C or D)

A dart is randomly thrown at a square dartboard. Write the probabilities as simplified fractions.

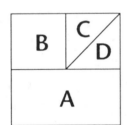 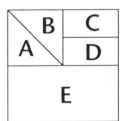

13. P(A)
14. P(D or C)
15. P(D)
16. P(~D)

17. P(E)
18. P(A or B)
19. P(A)
20. P(C or D)
21. P(D)
22. P(A or D)

Slips of paper are numbered from 1–25 and placed in a hat. One strip is drawn at random. Write each of the probabilities as a fraction and then as a percent.

23. P(even)
24. P(more than 20)
25. P(even, more than 20)
26. P(less than 5 or more than 20)
27. P(prime, less than 10)
28. P(multiple of 5)

Stella had 20¢ in her pocket.

29. List all of the possible ways she could have had 20¢.
30. What is the least number of coins she might have?
31. What is the greatest number of coins she might have?

PROBABILITY
Lesson Plan 8

QUIZ

None

OBJECTIVES

Students will perform an experiment to show that mathematics can accurately predict outcomes in real-world situations.

MATERIALS NEEDED

1. A copy of Experiment 1 for each student (transparency also)
2. Either a spinner or a bent paper clip for each pair of students
3. Problem Set 8

CLASS ACTIVITIES

Give each student a copy of Experiment 1. Pair the students up and designate one of them to be the *spinner* and the other to be the *recorder*. Give the *spinner* a paper clip that is bent so it can be used as a spinner. The *spinner* uses a pencil to anchor the looped end of the paper clip at the center of the circle and then flips it. Using tally marks, the *recorder* indicates which region of the circle the paper clip points to. Repeat 25 times. The *spinner* should flip each time from where the paper clip stopped. Starting the paper clip at the same place each time can lead to invalid data. The *spinner* should copy the *recorder's* data on his or her worksheet. After each pair is done, the teacher should compile the class total on the board. Discuss the fact that the class total (because of the large number of trials) should be very close to the theoretical probability.

Problem Set 8 contains no new material.

ASSIGNMENT

Problem Set 8

EXPERIMENT 1

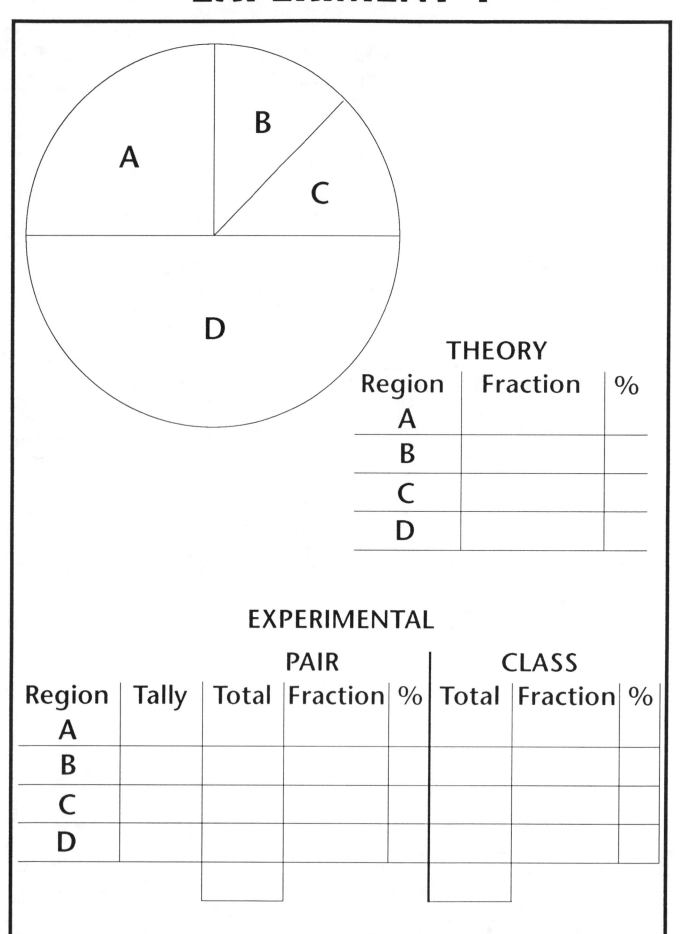

THEORY

Region	Fraction	%
A		
B		
C		
D		

EXPERIMENTAL

Region	Tally	PAIR Total	Fraction	%	CLASS Total	Fraction	%
A							
B							
C							
D							

PROBABILITY
Problem Set 8

Write the dartboard probabilities as fractions.

1. P(D)
2. P(B or C)
3. P(C)
4. P(~C)
5. P(A or D)
6. P(A or B)

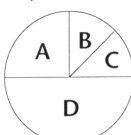

7. P(A or B)
8. P(C)
9. P(A)
10. P(~A)

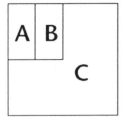

A card is drawn at random. Write the following probabilities as simplified fractions.

11. P(A)
12. P(vowel)
13. P(striped)
14. P(striped and vowel)
15. P(striped or vowel)
16. P(U)
17. What hero's name can be spelled using the cards?
18. The probability of having an accident is 2%. What is the probability of not having an accident?
19. The probability of getting struck by a unicycle is 3/50,000. What is the probability of not getting struck?

GUINNESS RECORD®: The greatest number of words typed in an hour on an electric typewriter is 9,316 by Margaret Hamma in Brooklyn, New York on June 20, 1941.

20. How many words did she type per minute?
21. About how long would it take her to type a 500 word theme?

BRAIN BUSTER: Freeda is between 10 and 30 years old. This year her age is a multiple of 7 and next year her age will be a multiple of 11. How old is Freeda?

PROBABILITY
Lesson Plan 9

QUIZ   Quiz 4

OBJECTIVES   Given a table of data, students will be able to interpret the
                        information in order to determine various probabilities.

MATERIALS NEEDED   1. Quiz 4
                        2. Transparency 9
                        3. Problem Set 9

CLASS ACTIVITIES   Spending time going over the quiz is worthwhile.
                        Transparency 9 introduces the new material.

ASSIGNMENT   Problem Set 9

PROBABILITY QUIZ 4

Name _____

A dart is thrown randomly at each dartboard.
Write the probabilities as reduced fractions.

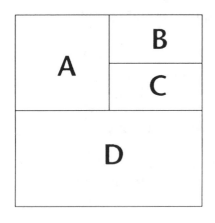
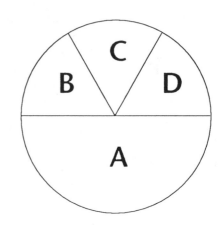

1. P(A) = _____

2. P(B or C) = _____

3. P(C) = _____

4. P(~C) = _____

5. P(A) = _____

6. P(C) = _____

7. P(~C) = _____

8. P(A or B) = _____

9. The probability of "forgetting" is 45/46.
   What is the probability of "remembering"?
   _____

10. Write 75% as a simplified fraction. _____

PROBABILITY 9

Michele Jordache has a large collection of basketball shoes. The table below shows how many of each type she has.

	Red	White
Hi Top	16	22
Lo Cut	9	3

1. How many pairs does she have? _____

2. How many pairs are red? _____

3. How many pairs are Lo Cut? _____

Before each game, she picks a pair at random.

4. P(white) =

5. P(lo cut) =

6. P(white and lo cut) =

7. P(white or lo cut) =

PROBABILITY
Problem Set 9

The table shows the distribution of students in a mathematics class.

1. How many students are in the class?
2. How many males are in the class?

	Male	Female
Juniors	7	16
Seniors	12	15

A student is chosen from the chart at random. Write the probabilities as fractions and percents.

3. P(male)   4. P(~male)   5. P(junior)   6. P(male or junior)
7. P(male and junior)

The table gives information about how students get to school. Some information is missing, but it can be determined.

8. How many males ride?
9. How many students walk?
10. How many students are in the school?

	Ride	Walk
Male	?	32
Female	38	46
Total	122	?

A student is chosen at random. Write the probabilities as fractions and as percents.

11. P(rides)   12. P(female)   13. P(male)   14. P(male or rides)
15. P(female and walks)

A dart is thrown at the dartboard. Write the probabilities as simplified fractions.

16. P(A)
17. P(D or E)
18. P(E)
19. P(A or C)
20. P(~C)

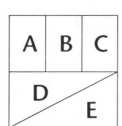

21. P(A)
22. P(D or E)
23. P(E)
24. P(A or C)
25. P(~E)

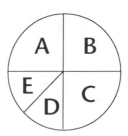

Three Flavors Yogurt Shoppe offers vanilla, strawberry, and chocolate yogurt and three toppings: peanuts, M&M's, and fudge.

26. Make a list of the nine types of cones that can be ordered. Your columns should be:

        Yogurt   *Topping*

PROBABILITY
Lesson Plan 10

QUIZ   None

OBJECTIVES   This lesson presents no new material. It is a review lesson in
                    preparation for the Unit Test.

MATERIALS NEEDED   Problem Set 10

CLASS ACTIVITIES   The students should be given about 25 minutes to work through
                    Problem Set 10. The rest of the period should be spent going over
                    the problems.

ASSIGNMENT   Problem Set 10

PROBABILITY
Problem Set 10

Write as percents.

1. 45/100   2. 68/200   3. 3/25   4. 9/50   5. 7/10   6. 15/300

Write as simplified fractions.

7. 50%   8. 75%   9. 110%   10. 44%   11. 2%   12. 64%

Estimate as percents.

13. 35/49   14. 8/9   15. 29/33   16. 188/199   17. 12/19   18. 4/11

19. The probability of Tarzan missing the rope was 4%. What was the probability that he would grab the rope?

20. The probability of being late was 13/20. What was the probability of being on time?

A card is drawn at random. Write the probabilities as simplified fractions.

21. P(E)
22. P(~E)
23. P(Q)
24. P(striped)
25. P(striped or E)
26. P(striped and E)
27. P(vowel or consonant)

PROBABILITY
Problem Set 10

A dart is thrown at random at the dartboards. Write the probabilities as simplified fractions.

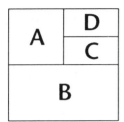
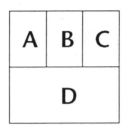
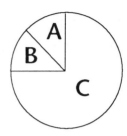

28. P(A)

29. P(C or D)

30. P(D)

31. P(A or B)

32. P(A)

33. P(A or D)

34. P(~A)

35. P(A or B)

36. P(A)

37. P(C)

38. P(A or C)

39. P(~B)

The table gives the results of a survey of who buys yogurt at the *Three Flavors Yogurt Shoppe.*

Use this table for problems 40–45.

	Male	Female
50 and under	23	19
over 50	45	13

40. How many people were surveyed?

41. How many people over 50 were surveyed?

A customer from the survey is chosen at random. Write the following probabilities as simplified fractions.

42. P(male)   43. P(over 50)   44. P(male or over 50)

45. P(male, over 50)

PROBABILITY
Test 1

Do not write on this test. Do all work on scratch paper and put the answers in the appropriate space on the answer sheet.

Write each of the following as a percent.

1. 57/100   2. 19/25   3. 19/20   4. 53/50   5. 78/200   6. 78/300

Estimate as percents.

7. 47/103   8. 8/9   9. 11/49   10. 45/501   11. 27/26   12. 28/32

Write as simplified fractions.

13. 90%   14. 75%   15. 125%   16. 13%   17. 100%

18. If P(A) = 70%, what is P(~A)?

19. If the probability of Elvis being alive is 3/10,000, what is the probability that Elvis is dead?

20. The probability of rolling two sixes in a row with a die is about 3%. What is the probability of not rolling two sixes?

A card is drawn at random from the cards. Write the probabilities as simplified fractions.

   T   E M A T   C

| M | A | T | H | E | M | A | T | I | C | S |

21. P(M)   22. P(V)   23. P(comes before Y in the alphabet)   24. P(striped)
25. P(vowel)   26. P(striped and vowel)   27. P(striped or vowel)   28. P(~M)

An experiment was run in which the spinner shown was spun. The table shows the results.

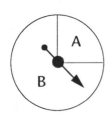

Region	Tally
A	ⅢⅡ II
B	ⅢⅡ ⅢⅡ ⅢⅡ III

29. How many times was the spinner spun?

30. How many times did it stop in A?

31. What percent of the time did it stop in A?

32. According to *theory*, what percent of the time should it stop in A?

PROBABILITY
Test 1

A dart is thrown at the dartboard. Write the probabilities as simplified fractions.

33. P(A)
34. P(B or C)
35. P(~A)
36. P(B)
37. P(A or B)

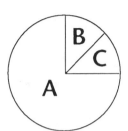

38. P(A)
39. P(B or C or D)
40. P(B)
41. P(B or C)
42. P(A or B)

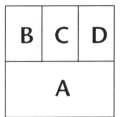

Egore has a large number of relatives that he classifies according to the chart.

	Aunts	Uncles
Good	6	7
Bad	12	8
Ugly	3	11

43. How many relatives does Egore have?
44. How many aunts does Egore have?
45. How many ugly relatives does Egore have?

Egore randomly chooses one of these relatives to take to lunch. Write the following probabilities as fractions.

46. P(aunt)   47. P(ugly)   48. P(ugly, an aunt)   49. P(good or bad)
50. P(good or ugly or an aunt)   51. P(bad and ugly)   52. P(aunt or uncle)

Theo has some coins totaling 17¢.

53–58. List the six possible combinations of coins.

59. What is the greatest number of coins he could have had?
60. What is the least number of coins he could have had?
61. Could he have five coins?
62. Could he have six coins?

PROBABILITY

Test 1 Answer Sheet

Name _____

1. _____   22. _____   43. _____

2. _____   23. _____   44. _____

3. _____   24. _____   45. _____

4. _____   25. _____   46. _____

5. _____   26. _____   47. _____

6. _____   27. _____   48. _____

7. _____   28. _____   49. _____

8. _____   29. _____   50. _____

9. _____   30. _____   51. _____

10. _____   31. _____   52. _____

11. _____   32. _____   *Dimes   Nickels   Pennies*

12. _____   33. _____   53. _____

13. _____   34. _____   54. _____

14. _____   35. _____   55. _____

15. _____   36. _____   56. _____

16. _____   37. _____   57. _____

17. _____   38. _____   58. _____

18. _____   39. _____   59. _____

19. _____   40. _____   60. _____

20. _____   41. _____   61. _____

21. _____   42. _____   62. _____

PROBABILITY
Lesson Plan 11

QUIZ   None

OBJECTIVES   1. Given a two-stage experiment, students will be able to
                            construct a table listing the outcomes and then use the table to
                            determine probabilities.

                        2. Given a decimal representation of a number, students will be
                            able to write the decimal to the nearest 100th and then as
                            a percent.

MATERIALS NEEDED   1. Transparencies 11A and 11B

                        2. Problem Set 11

                        3. A pair of dice (optional)

CLASS ACTIVITIES   Transparencies 11A and 11B cover the new material. After the table
                        is complete on Transparency 11A, questions similar to those in
                        Problem Set 11 should be posed. Point out that the fractions total
                        36/36 or 1. Make sure that the students copy Transparency 11A into
                        their notes as it is needed in working Problem Set 11. Also,
                        announce to the students that they will need a calculator during
                        this unit.

ASSIGNMENT   Problem Set 11

PROBABILITY 11A

A pair of dice are rolled. The numbers on the top are added. List all possible outcomes and determine their probabilities.

+	1	2	3	4	5	6
1						
2						
3						
4						
5						
6						

P(2) =   P(6) =   P(10) =

P(3) =   P(7) =   P(11) =

P(4) =   P(8) =   P(12) =

P(5) =   P(9) =

PROBABILITY 11B

Round to the nearest 100th, then write as a percent.

1. .56487

2. .5681908

3. .3

4. .349364

5. .29703

6. .03963

7. .0073298

PROBABILITY
Problem Set 11

To answer problems 1– 10, refer to the table in your notes showing the outcomes of rolling a pair of dice.

1.   What sum is most likely to appear?   2.   What sums are least likely to appear?

3.   The probability of rolling a 5 is the same as the probability of rolling what other number?

Write answers to the following as simplified fractions.

4.   P(2 or 12)   5.   P(more than 8)   6.   P(less than 4)

7.   P(even, more than 8)   8.   P(even or more than 8)   9.   P(multiple of 5)

10. P(divisible by 4)

Round to the nearest 100th and then write as a percent.

11. .4523   12. .4669143 13. .0312348 14. .25   15. .8   16. .496112

17. .006921   18. 1.13674

Write as percents.

19. 1/4   20. 17/25   21. 3/4   22. 1/2

Approximate as percents.

23. 1/3   24. 13/21   25. 5/9   26. 2/49

The spinner is spun twice and the numbers that result are multiplied.

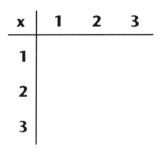

27. Copy and complete the table of possible outcomes.

Write answers to the following as simplified fractions.

28. P(1)   29. P(5)   30. P(less than 10)   31. P(even)   32. P(~even)

33. P(even, less than 5)   34. P(even or less than 5)

BRAIN BUSTER: Draw the path from Start to End.

START   END

5   10

÷11   +7

x 6   +3

PROBABILITY
Lesson Plan 12

QUIZ   None

OBJECTIVES   Given a fraction, students will use calculators to convert it to a
                        whole percent.

MATERIALS NEEDED   1. Transparencies 12A and 12B
                        2. Problem Set 12
                        3. Tetrahedral die (optional)

CLASS ACTIVITIES   Transparency 12A gives students more practice in constructing
                        tables listing all outcomes. Transparency 12B introduces students to
                        using a calculator to convert fractions to percents. Emphasize that
                        calculators should only be used when the problem cannot be done
                        mentally. The C/M column indicates whether a *calculator* or *mental*
                        mathematics should be used. The new column should contain a
                        fraction with a denominator of 100 or a decimal obtained from a
                        calculator.

ASSIGNMENT   Problem Set 12

PROBABILITY 12A

Two tetrahedral die are rolled. The sum of their down faces is computed.

+	1	2	3	4
1				
2				
3				
4				

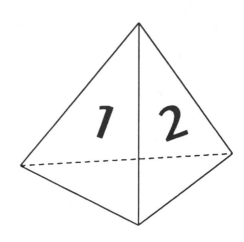

1. P(3)

2. P(~3)

3. P(10)

4. P(less than 12)

5. P(even)

6. P(even and more than 4)

7. P(even or more than 4)

PROBABILITY 12B

Write each of the following as whole percents.

	C/M	new	%
1.	14/50		
2.	14/23		
3.	13/53		
4.	3/10		
5.	11/63		
6.	7/25		
7.	16/123		
8.	150/200		
9.	150/190		
10.	45/38		

PROBABILITY
Problem Set 12

Write the following as whole percents. Do not use a calculator.

1. 34/50   2. 1/2   3. 11/25   4. .89   5. .327   6. .367
7. .8993   8. .7   9. 1.12

Use a calculator to write as whole percents.

10. 13/47   11. 57/89   12. 134/121   13. 83/87   14. 3/312   15. 1/16

The spinner is spun and a standard die is tossed. Your score is the sum of the two numbers that appear.

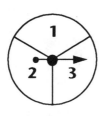

spinner

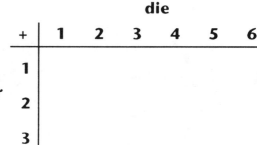

16. Copy and complete the table.
17. What sum is most likely to occur?

Write the probabilities as fractions and whole percents. Use a calculator.

18. P(2)   19. P(4 or 7)   20. P(less than 7)
21. P(even)   22. P(even, less than 7)   23. P(even or less than 7)

Darts are thrown at the hexagonal dartboards. Write the probabilities as simplified fractions.

24. P(A)
25. P(~A)
26. P(B or C)
27. P(A or C or D)

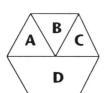

28. P(B)
29. P(A or C)
30. P(B or C)
31. P(~A)

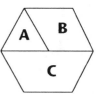

GUINNESS RECORD®: The largest snake ever held in captivity was a reticulated python named Colossus. It measured 342 inches long and weighed 5120 ounces.

32. How long was it in feet?
33. How much did it weigh in pounds?

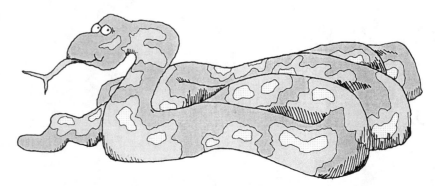

PROBABILITY
Lesson Plan 13

QUIZ

Quiz 5

OBJECTIVES

There is no new material in this lesson.

MATERIALS NEEDED

1. Quiz 5
2. Experiment 2. A transparency of this will also be helpful.
3. A pair of dice and a styrofoam cup to be used as a shaker. One set for every two students.

CLASS ACTIVITIES

Give each pair of students a pair of dice and a styrofoam cup. Have them shake the dice and tally their results for a large number of trials. Complete the table for individual pairs of students and for the class. The theoretical probabilities (%) come from the notes from Lesson 11. This activity should take approximately 30 minutes.

ASSIGNMENT

None

PROBABILITY QUIZ 5

Name _____

Round as necessary and then write as a whole percent.

1. .347 = _____   3. .8 = _____

2. .5919 = _____   4. 1.197 = _____

Two tetrahedral die are rolled and the numbers "down" are multiplied. Complete the table.

5.

x	1	2	3	4
1				
2				
3		12		
4				

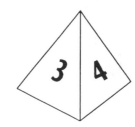 

Write the probabilities as simplified fractions.

6. P(odd) = _____

7. P(more than 7) = _____

8. P(more than 7 and even) = _____

9. P(more than 7 or even) = _____

EXPERIMENT 2

Rolling Two Dice

THEORETICAL PROBABILITY		Tally	PAIR		CLASS	
Outcome	%		Total	%	Total	%
2						
3						
4						
5						
6						
7						
8						
9						
10						
11						
12						

PROBABILITY
Lesson Plan 14

QUIZ

None

OBJECTIVES

Given a word problem of the type "*a* of the *b* had a certain characteristic," students will be able to determine the percent having the characteristic.

MATERIALS NEEDED

1. Transparency 14A and B
2. Problem Set 14
3. A misspotted pair of dice (optional)

CLASS ACTIVITIES

Transparency 14A can be used to explain how cheating can take place in certain dice games. In Craps, the customer has to roll his number (his "point," determined by rolling the dice) before rolling a 7. Suppose the customer has an even number as his goal. The person running the game substitutes the misspotted dice, one marked 1, 3 and 5, and the other marked 2, 4, and 6 into the game. (It is only possible to see three faces of a die at one time). It is then impossible for the customer to make his point. Transparency 14B has examples of percent problems. Note that some are calculator problems and some are mental problems. A percent line is useful in determining the reasonableness of one's answer. The percent line for Example 3 is below.

From the diagram, it is obvious that the answer is a number between 0 and 25.

ASSIGNMENT

Problem Set 14

PROBABILITY 14A

The Misspotted Dice

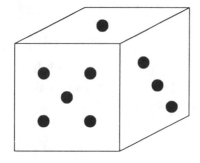 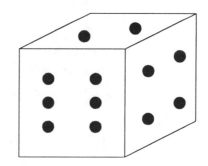

Even Die

	2	2	4	4	6	6
1						
1						
3						
3						
5						
5						

Odd Die

PROBABILITY 14B

Percent Problems

Of 100 students surveyed, 87 said that they owned a CD player. What percent is that?

Of 50 students surveyed, 19 said that they had a soft drink for breakfast. What percent is that?

Of 150 students surveyed, 23 said that they walked to school. What percent is that?

Of 64 students interviewed, 57 said that they believed in The Loch Ness Monster. What percent is that?

PROBABILITY
Problem Set 14

One hundred teenagers were interviewed.

1. Thirty-four liked spinach. What percent liked spinach?
2. Eighty-three liked carrots. What percent liked carrots?
3. Seven liked broccoli. What percent liked broccoli?

Twenty-five first graders were interviewed.

4. Twenty-two watched Saturday morning cartoons. What percent watched the cartoons?
5. Three had flown on an airplane. What percent had flown?
6. Seventeen had been to Disney World. What percent had been to Disney World?

Seventy-seven adults were surveyed concerning their preference of pizza crusts.

7. Forty preferred thin. What percent preferred thin?
8. Thirty preferred deep dish. What percent preferred deep dish?
9. Seven had no preference. What percent had no preference?

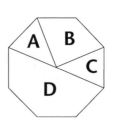

A roll of toilet paper was tossed into the air 6,578 times and landed on its end 97 times.

10. What percent of the time did it land on its end?
11. How many times did it not land on its end?
12. What percent of the time did it not land on its end?

The spinner is spun twice. The score is the sum of the two spins.

13. Copy and complete the table listing the outcomes.
14. What sum is most likely to appear?

Write the probabilities as whole percents.

15. P(10)
16. P(9)
17. P(odd)
18. P(more than 6)
19. P(more than 6 or even)
20. P(more than 6 and even)

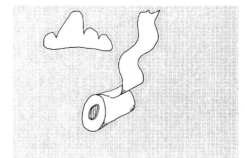

+	1	2	3	5
1			4	
2				
3				
5				

A dart is thrown at the octagonal dartboard.
Write each of the probabilities as simplified fractions.

21. P(A)
22. P(B)
23. P(~C)
24. P(A or C)
25. P(A or B)
26. P(A or B or D)
27. The probability of fog is 3%. What is the probability of no fog?

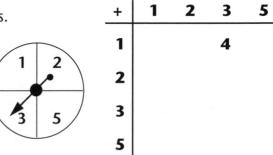

PROBABILITY
Lesson Plan 15

QUIZ

Quiz 6

OBJECTIVES

Students will be able to draw a tree diagram listing the outcomes in a multi-stage event.

MATERIALS NEEDED

1. Quiz 6
2. Transparencies 15A–C
3. Problem Set 15

CLASS ACTIVITIES

Transparency 15A is a multiple-choice "quiz." Since the students will be guessing on each question, the probability of a student getting all three correct is 1/24. You might offer a reward—a free late, a pencil, 5 bonus points on the next test—to any student who gets all three correct. After the quiz is given, show the three-stage, 24-branch tree and list all of the possible outcomes. The correct answers are B, C, and A. Transparency 15C gives examples that are similar to those found in the homework. In drawing trees, it is important that all branches from corresponding stages end on the same vertical line. Drawing the lines is helpful.

ASSIGNMENT

Problem Set 15

PROBABILITY QUIZ 6

Name _____

Write as whole percents.

1. 7/10 = _____   3. 142/200 = _____

2. .7 = _____   4. 17/43 = _____

The spinner is spun twice. Your score is the larger of the two numbers. For example, 3 then 2 gives 3, and 2 then 2 gives 2.

	2	3	4
2			
3			
4			

5. Complete the table.

Write as simplified fractions.

6. P(2) = _____

7. P(4) = _____

8. P(~4) = _____

9. P(more than 2) = _____

10. P(more than 2 or even) = _____

PROBABILITY 15A

According to a scientific poll of teenagers conducted by the World Almanac in 1989

1. their favorite book was:

   a. Pet Semetary

   b. The Outsiders

   c. The Bible

2. their favorite television show was:

   a. ALF

   b. The Wonder Years

   c. Roseanne

   d. The Cosby Show

3. their "hero" was:

   a. Michael Jordan

   b. Tom Cruise

PROBABILITY 15B

Multiple choice quiz answers

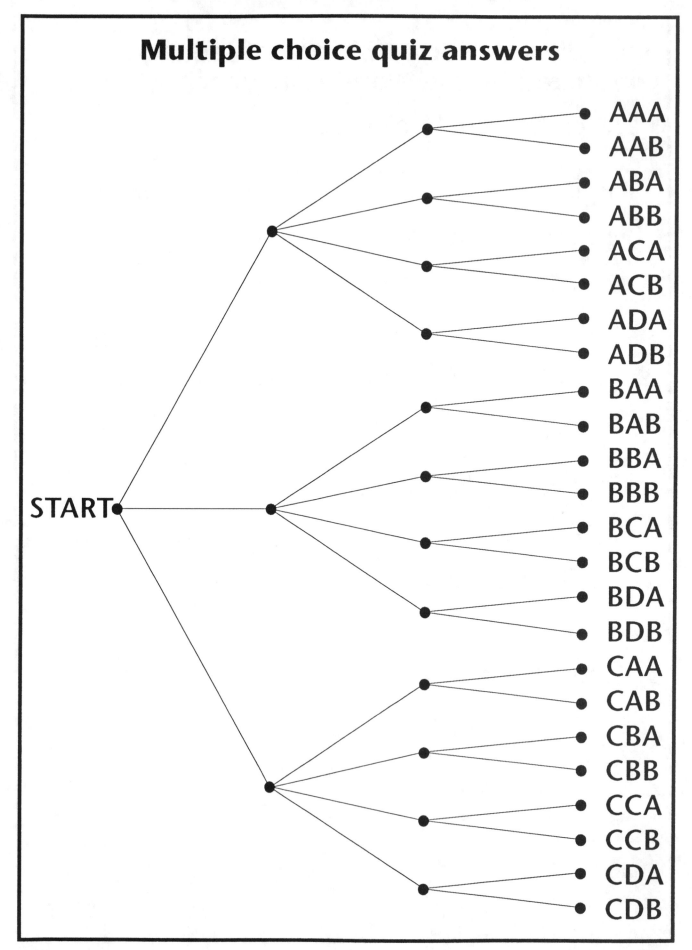

PROBABILITY 15C

Tree Examples

Example 1: The spinner is spun and a coin is tossed. List all outcomes.

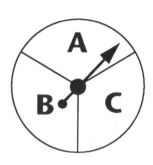

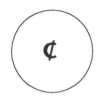

Example 2: A tetrahedral die is tossed and the spinner is spun. List all outcomes.

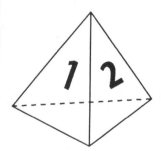

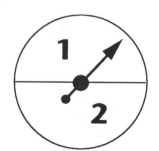

PROBABILITY
Problem Set 15

For problems 1–4, draw the tree diagrams and list all outcomes at the end of the branches.

1. Spin the spinner and toss the coin. This will give a 2-stage, 4-branch tree.

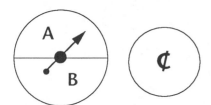

2. Spin the spinner and toss the coin. This will give a 2-stage, 8-branch tree.

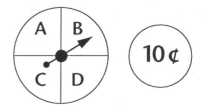

3. Choose a card and toss the coin.

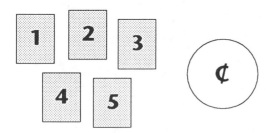

4. Toss a penny, then a dime, then a quarter. This will give a 3-stage, 8-branch tree.

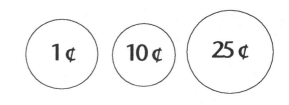

Write as whole percents. You should use the calculator only twice.

5. 13/50   6. 13/37   7. .64   8. .9   9. 42/300   10. 21/20

11. .2694   12. 1.05   13. 3456/7890

A survey was taken of 200 students.

14. Forty are left-handed. What percent is that?
15. One hundred and twenty wear glasses. What percent is that?
16. Two hundred like popcorn. What percent is that?

A survey was taken of 345 people attending a wrestling match.

17. This was the first match that 66 had attended. What percent is that?
18. How many had attended other matches?
19. What percent had attended other matches?
20. Three hundred people believed that the matches were real. What percent is that?
21. How many people thought that the matches were fake?
22. What percent believed that the matches were fake?

PROBABILITY
Lesson Plan 16

QUIZ   None

OBJECTIVES   Given a multi-stage event, students will be able to draw the tree
                       diagram and calculate probabilities from it.

MATERIALS NEEDED   1. Transparency 16A and 16B
                       2. Problem Set 16

CLASS ACTIVITIES   Transparency 16A reviews the percent line concept. The task is to fill
                       in the numbers above the percents. None of these are calculator
                       problems. Transparency 16B introduces students to using trees to
                       calculate probabilities in multi-stage events. Note that P(A,B) means
                       A *then* B. P(B,A) means B *then* A. Also, question 6 can be done either
                       by looking at the picture or by using the tree.

ASSIGNMENT   Problem Set 16

PROBABILITY 16A

Percent Lines

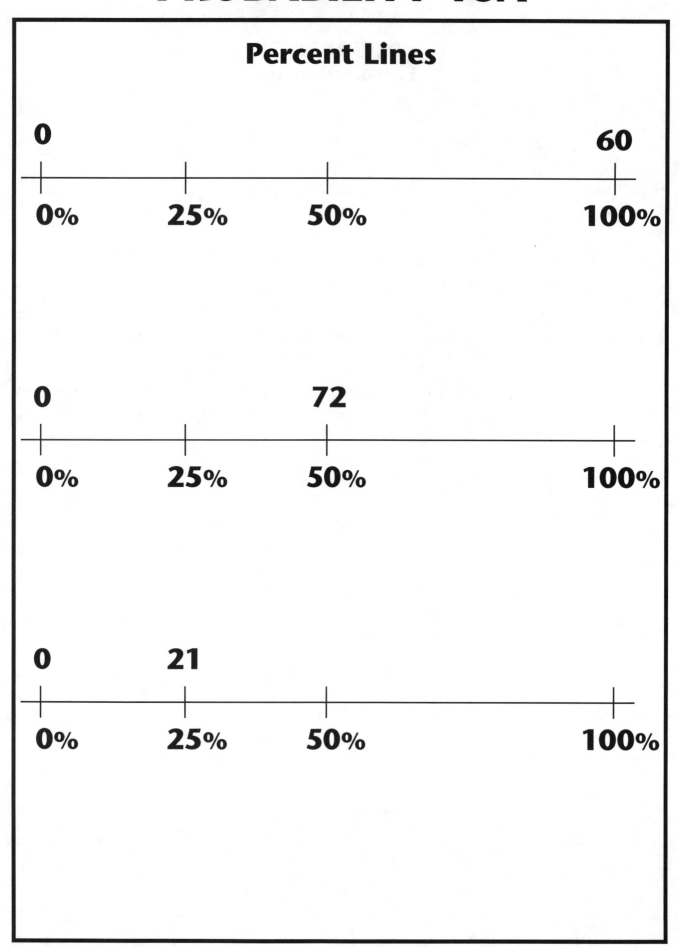

0   60

0%   25%   50%   100%

0   72

0%   25%   50%   100%

0   21

0%   25%   50%   100%

PROBABILITY 16B

The two spinners are spun.

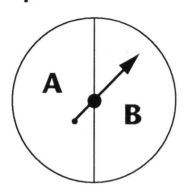 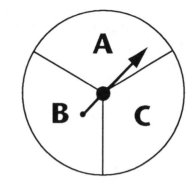

1. Draw a tree showing all outcomes.

Write as simplified fractions.

2. P(A,C) =

3. P(C,A) =

4. P(letters are the same) =

5. P(letters are different) =

6. P(second spinner shows A) =

PROBABILITY
Problem Set 16

Spinner 1 and then spinner 2 is spun.

1. Draw a two-stage, eight-branch tree diagram listing all possible outcomes.

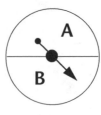
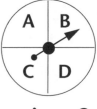

spinner 1   **spinner 2**

Write the following as simplified fractions.

2. P(A,A)   3. P(A,C)   4. P(C,A)   5. P(A on first spin)

6. P(letters match)   7. P(letters different) 8. P(B on second spin)

Elvira has won a contest. To determine how much she has won, she spins a spinner twice. Her take is the sum of the two spins.

9. Draw a tree diagram listing the nine possible outcomes.
10. What is her largest possible take?
11. What is her smallest possible take?

Write the following as simplified fractions.

12. P($1,$5)   13. P($5,$1)   14. P(win $6)   15. P(win more than $5)

16. P($20,$20)   17. P(win less than $50)

Write the following as whole percents. Use a calculator twice.

18. 4762/7638   19. 134/121   20. 17/20   21. .799352   22. .9

23. .09   24. 230/200   25. .132   26. .0073

Draw a percent line to work the following four problems. Eighty customers were surveyed.

27. Forty liked Gloomios. What percent liked Gloomios?
28. Eight liked White Ties. What percent liked White Ties?
29. None liked Gooey Bears. What percent is that?
30. Fifty-seven liked Flax Flakes. What percent is that?

BRAIN BUSTER: Use 19 coins to make a $1.00. You can only use two different kind of coins.

PROBABILITY
Lesson Plan 17

QUIZ   Quiz 7

OBJECTIVES   Students will verify that tree diagrams do predict the outcome of a multi-stage experiment. Students will apply the *Guess and Test* problem-solving strategy to appropriate problems.

MATERIALS NEEDED   1. Quiz 7
                        2. Experiment 3 (a transparency also)
                        3. A styrofoam cup and three coins for each pair of students.
                        4. Transparency 17

CLASS ACTIVITIES   Each pair of students needs a styrofoam cup (used as a shaker) and three coins. Have students shake the coins an "appropriate" number of times. Each pair should record their results. Summarize results for the class. Then draw the three-stage tree and compute the theoretical probabilities. Note that we are not interested in the order of H's and T's, simply the number. Answer the questions at the bottom of the paper. The back of the paper can be used to work the *Guess and Test* problems from Transparency 17.

ASSIGNMENT   None

PROBABILITY QUIZ 7

Name _____

Eighty people were interviewed.

1. Twenty were left-handed. What percent were left-handed? _____

2. Sixty-seven liked Shredded Oats. What percent is that? _____

The spinner is spun twice. Your score is 5 more than the smaller number. For example, 3 followed by 4 gives you a score of 8.

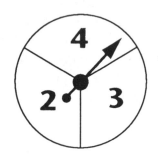

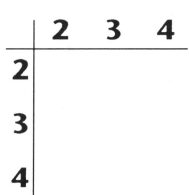

	2	3	4
2			
3			
4			

3. Complete the table.

Write as simplified fractions.

4. P(8) = _____

5. P(less than 8) = _____

6. P(odd) = _____

7. P(odd and less than 8) = _____

EXPERIMENT 3

Tossing Three Coins

Outcome	Tally	PAIR		CLASS		THEORY
		Total	%	Total	%	
3H, 0T						
2H, 1T						
1H, 2T						
0H, 3T						

Draw a three-stage, eight-branch tree for this experiment and list the outcomes.

START ●

Use the tree to answer the following. Write the probabilities as fractions and whole percents.

1. P(all heads or all tails) =

2. P(at least two heads) =

3. P(heads on the first toss) =

4. P(heads on the third toss) =

5. P(at least one tail) =

PROBABILITY 17

Guess and Test

Sayeth Egore… "If you triple my age, subtract 4, and divide by 2, you'll get 13." How old is Egore?

Guess	x 3	– 4	÷ 2

Sayeth Obeseth… "In three days I ate 81 pizzas. On each day I ate 10 more pizzas than on the previous day." How many pizzas did Obeseth eat on the first day?

Guess	Day 2	Day 3	Total

PROBABILITY
Lesson Plan 18

QUIZ   None

OBJECTIVES   No new ideas are introduced in this lesson.

MATERIALS NEEDED   1. Transparency 18A and 18B
                        2. Problem Set 18

CLASS ACTIVITIES   Transparency 18A gives two more examples of *Guess and Test*
                        problems. Transparency 18B gives a slightly different twist to a
                        percent problem.

ASSIGNMENT   Problem Set 18

PROBABILITY 18A

Guess and Test

In four days, Firth made 50¢ by picking fleas off dogs. Each day he made 4¢ less than the day before. How much did he make the first day?

Guess	Day 2	Day 3	Day 4

In three hours, Gertrude built 70 bridges. Each hour she made half as many as the hour before. How many bridges did she make the first hour?

Guess	Hour 2	Hour 3

PROBABILITY 18B

Percent

The price of a new boomerang was $453. However, Fenwick was able to purchase it for only $376.

1. What percent of the original price did Fenwick pay?

2. How much money did Fenwick save?

3. What percent of the original price did Fenwick save?

PROBABILITY
Problem Set 18

Ivan sayeth, "Multiply the amount of money that I have by 3, add one, and divide by 2. The result is 20¢."

1. Use the *Guess and Test* procedure to find out how much money Ivan has.

2. Use the *Make a List* procedure to list four combinations of coins that Ivan could have.

Abagail sayeth, "Take my age, divide it by 2, add 3, divide by 5 and subtract 5. The result is 0."

3. Use the *Guess and Test* procedure to determine Abagail's age.

4. On a test, Bedelia got 37 out of 50 correct. Determine the percent that Bedelia got correct.

5. Seventeen of 39 people surveyed ate Wheat Meal for breakfast. Determine the percent.

6. Forty-six of 46 people surveyed preferred Preparation M. What percent of the people preferred Preparation M?

7. Bevis paid $450 for a car that originally cost $900. What percent of the original price did he pay?

8. A harmonica originally cost $60. Abishag was able to purchase it for $54. What percent of the original price did Abishag pay?

9. Because an item was on sale, Egbert only had to pay 65% of the original price. What percent did Egbert save?

A card is drawn from Deck One and then one is drawn from Deck Two.

 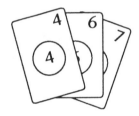

10. Draw a two-stage, nine-branch tree and list all outcomes.

Write the following probabilities as reduced fractions.

11. P(4,7)

12. P(total is 11)

13. P(4 on first draw)

14. P(cards match)

15. P(second card is even)

16. P(total is a prime number)

17. Four coins are tossed. If a tree diagram were drawn, how many branches would it have?

GUINNESS RECORD®: Theodore Coombs of Hermosa Beach, California roller skated 5,193 miles from Los Angeles to New York City and back to Yates Center, KS in 116 days during the summer of 1979.

18. To the nearest mile, how many miles did he average each day?

19. If he skated 9 hours each day, how many miles did he average each hour?

PROBABILITY
Lesson Plan 19

QUIZ   Quiz 8

OBJECTIVES   This is a review lesson in preparation for the test on Unit Two.

MATERIALS NEEDED   1. Quiz 8
                        2. Problem Set 19

CLASS ACTIVITIES   Pass out Problem Set 19 at the beginning of class. Either work
                        through the problems with the students or let them work for about
                        30 minutes and then check their work.

ASSIGNMENT   Problem Set 19

PROBABILITY QUIZ 8

Name _____

The spinner is spun and then a card is drawn.

1. Draw the two-stage, eight-branch tree and list all outcomes.

Write all probabilities as simplified fractions.

2. P(G,M) = _____   3. P(letters match) = _____

4. P(G on spin) = _____

5. P(letters don't match) = _____

6. P(letters are both vowels) = _____

7. P(M,G) = _____

PROBABILITY
Problem Set 19

1. Sayeth Delwyn, "Add 5 to my IQ, multiply by 2, add 20 and divide by 4. The result is 50." Use the *Guess and Test* procedure to determine Delwyn's IQ.

2. Jessica did 62 fingersnaps in four minutes. Each minute she did one more than the previous minute. How many fingersnaps did Jessica do the first minute?

3. Delwyn made $28 in three days. Each day he made half as much as he did the day before. How much did Delwyn make the first day?

4. Damarias won 17 out of 23 chariot races. What percent did he win?

Twenty-five people were interviewed.

5. Three liked Valley Dew. What percent is that?

6. Twenty-four liked Mella Yella. What percent is that?

7. Twenty-five liked Himshey bars. What percent is that?

The sticker price on a new car was $12,995 but Ainslie was able to purchase it for $11,500.

8. What percent of the sticker price did Ainslie pay?

9. How much money did Ainslie save?

10. What percent of the sticker price did Ainslie save?

Emelia throws two darts at the dartboard. Her score is one more than the sum of the numbers. For example, 4 and 6 gives her a score of 11.

11. Copy and complete the table.

12. What is her highest possible score?

13. What is her most likely score?

	4	5	6
4	9		
5			
6			13

Write as simplified fractions.

14. P(even)   15. P(more than 11)   16. P(even, more than 11)

17. P(even or more than 11)

Draw the tree diagrams for the following experiments.

18. A nickel and then a dime are tossed (four branches).

19. A tetrahedral die labeled A, B, C, and D is tossed and a spinner labeled A and B is spun (eight branches).

BRAIN BUSTER: Find the missing number.

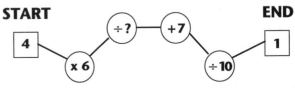

PROBABILITY
Test 2

Do not write on this test. Place all answers in the appropriate space on the answer sheet.

A regular (six faced) die and a tetrahedral (four faced) die are rolled. Your score is the sum of the up face on the regular die and the down face on the tetrahedral die.

+	1	2	3	4	5	6
1						
2						
3		5				
4						

1. Copy and complete the table.
2. What is the highest possible score?
3. Three scores are equally likely. What are they?
4. The probability of scoring three is the same as the probability of scoring _____.

Write the following as simplified fractions.

5. P(2)
6. P(even)
7. P(~even)
8. P(more than 7)
9. P(even and more than 7)
10. P(even or more than 7)
11. P(less than 12)
12. P(multiple of 5)

The spinner is spun twice. Your score is eight more than the smaller number. A one and a three would result in a score of nine.

13 Copy and complete the table.
14. What is the highest possible score?
15. What score is most likely?
16. What score is least likely?

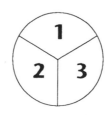

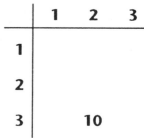

	1	2	3
1			
2			
3			10

Write as whole percents.

17. 1/2
18. 47/50
19. 47/68
20. .8
21. .08
22. .8632
23. .1973
24. 1.37131
25. 46/41

There are 30 players on the roster of the Kairo Kubs curling team.

26. Fifteen are right handed. What percent are right handed?
27. Twenty-three are bald. What percent are bald?
28. How many are not bald?
29. What percent are not bald?

PROBABILITY
Test 2

A porker was priced at $247. However, Emerson was able to purchase it for $183.

30. What percent of the regular price did Emerson have to pay?

31. How much money did Emerson save?

32. On the anthropology test, Brenda got 27 of the 35 questions correct. What percent did she get correct?

33. Of 64 people interviewed, none liked Oat Cryspies. What percent like Oat Cryspies?

34. Alvis wanted to look good for the big Math Test. He had five sweatshirts and seven pair of jeans. How many different outfits could he make?

A coin is tossed five times.

35. If a tree were drawn for this experiment, how many branches would it have?

36. What outcome would the top branch show?

Use the *Guess and Test* procedure in working the next two problems.

37. Egan made 45¢ in four days by washing cars. Each day she made twice as much as the previous day. How much did Egan make on the first day?

38. Agatha sayeth, "Begin with my age. Add 7, multiply by 5, divide by 25. You will end at 2." How old is Agatha?

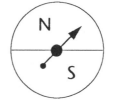

Spinner one and then Spinner two is spun.

39. Draw a two-stage, eight-branch tree for this experiment.

spinner 1   **spinner 2**

Write the following probabilities as simplified fractions.

40. P(N,W)   41. P(W,N)   42. P(letters match)

43. P(W on second spin)

A card is drawn from the deck, replaced, and a second is drawn.

44. Draw the two-stage, nine-branch tree for this experiment.

Write the following probabilities as simplified fractions.

45. P(1,3)   46. P(cards match)   47. P(total is 4)

48. P(total is even)   49. P(total is less than 8)

50. P(second card is more than first)

PROBABILITY

Test 2 Answer Sheet

Name _____

1.

+	1	2	3	4	5	6
1						
2						
3						
4						

2. _____

3. _____

4. _____

5. _____

6. _____

7. _____

8. _____

9. _____

10. _____

11. _____

12. _____

13.

	1	2	3
1			
2			
3			

14. _____

15. _____

16. _____

17. _____

18. _____

19. _____

20. _____

21. _____

22. _____

23. _____

24. _____

25. _____

26. _____

27. _____

28. _____

29. _____

30. _____

31. _____

32. _____

33. _____

34. _____

35. _____

36. _____

37. _____

38. _____

39.

40. _____

41. _____

42. _____

43. _____

44.

45. _____

46. _____

47. _____

48. _____

49. _____

50. _____

PROBABILITY
Lesson Plan 20

QUIZ

None

OBJECTIVES

In Experiment 4, students will learn that if the outcomes of an experiment are not equiprobable, one cannot simply look at the number of outcomes on the tree. Rather, the probability of each event must be written on each branch and then the probabilities multiplied to determine the probabilities of the outcomes.

MATERIALS NEEDED

1. Experiment 4 (Transparency also.)
2. Styrofoam cup with 2 white and 3 blue cubes for each pair of students.

CLASS ACTIVITIES

Give each pair of students a cup with the colored cubes. Have the students run the experiment a number of times. Emphasize that a trial consists of a draw, a replacement, and a second draw. Have each pair of students calculate their probabilities and the class probabilities. Do not calculate the theory. Draw the tree and calculate the probabilities without putting the probabilities on the branches. These probabilities (all 25%) will not agree with the experimental outcomes. The reason, the answer to the question at the bottom of the page, is that the individual probabilities are not equal: P(blue) = 3/5.

Turn the paper over and redraw the tree putting the probabilities on each branch. P(B,B) = 3/5 x 3/5 = 9/25 = 36%. Note that one's intuition says that B,B is the most likely outcome.

ASSIGNMENT

None

EXPERIMENT 4

Choosing colored cubes

Place 3 blue and 2 white cubes in a cup. Shake the cubes and draw one. Replace it and repeat the procedure. Record the result. Repeat the procedure until told to stop.

| | EXPERIMENT | | | | | THEORY |
| | PAIR | | | CLASS | | |
Outcome	Tally	Total	%	Total	%	%
B, B						
B, W						
W, B						
W, W						

Draw a two-stage tree for this experiment.

START

Why do the theoretical results not agree with our experimental results?

PROBABILITY
Lesson Plan 21

QUIZ   None

OBJECTIVES   Given a two-stage event in which the outcomes in a particular stage are not equiprobable, students will draw the tree and write the probabilities on the branches. The two stages are independent. Students will then multiply the branch fractions to calculate the various probabilities.

MATERIALS NEEDED   1. Transparency 21
                        2. Problem Set 21

CLASS ACTIVITIES   Work through at least two of the examples from Transparency 21. Calculate probabilities as fractions and percents.

                        The first example should look like this:

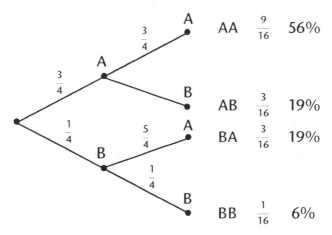

                        Appropriate discussion questions are:
                        1. Most likely outcome?
                        2. Least likely outcome?
                        3. Probability the darts match?
                        4. Sum of probabilities? 16/16 or 100%

ASSIGNMENT   Problem Set 21

PROBABILITY 21A

Trees—Unequal Outcomes

Two darts are thrown, one at each dartboard. Draw the tree, list outcomes and probabilities.

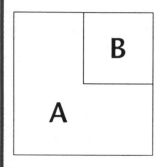

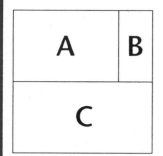

The spinner is spun twice. Draw the tree, list outcomes and probabilities.

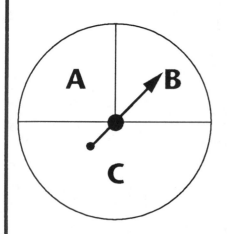

PROBABILITY 21B

Percents To Decimals

61% =

89% =

7% =

70% =

56.8% =

137% =

100% =

PROBABILITY
Problem Set 21

For each of the following spinners, the spinner is spun twice. For the dartboards, two darts are thrown. For problems 1–4, draw the two-stage trees, write the probabilities on the branches, and list all outcomes and their probabilities as simplified fractions.

1.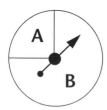
2.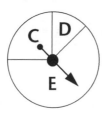
3.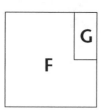
4.

There are 1800 students enrolled at Rahrah High School.

5. Nine hundred are girls. What percent are girls?
6. 465 eat lunch at school. What percent eat lunch at school?
7. 698 have jobs. What percent have jobs?

Write as decimals.

8. 56%   9. 38%   10. 100%   11. 28.7%   12. 6%   13. 156%

Alphonse attempted 39 shots but only made 19.

14. How many did Alphonse miss?
15. What percent did Alphonse make?
16. What percent did Alphonse miss?

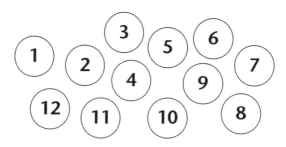

A jar contains disks numbered from 1–12. A disk is drawn at random. For each game write the probability of each player winning as a simplified fraction.

Game One
17. A wins if less than 8
18. B wins if more than 8
19. C wins otherwise

Game Two
20. A wins if more than 6
21. B wins if less than 4
22. C wins otherwise

Game Three
23. A wins if a multiple of 3
24. B wins if a multiple of 5
25. C wins otherwise

Game Four
26. A wins if 12
27. B wins if one digit
28. C wins otherwise

PROBABILITY
Lesson Plan 22

QUIZ

None

OBJECTIVES

Using the probability of an event occurring, students will be able to determine how many times the event should occur in a "nice number" of trials.

MATERIALS NEEDED

1. Transparency 22
2. Problem Set 22

CLASS ACTIVITIES

Complete the tree for the indicated problem, expressing the probabilities as both fractions and percents. Note that the number of trials is either a multiple of the denominator of the fraction or a number that can be obtained easily from 100.

ASSIGNMENT

Problem Set 22

PROBABILITY 22

The Unfair Coin

A coin is weighted so that P(H) = 5/9.
It is tossed twice.

1. The most likely outcome?

2. P(T,T) = _____

3. P(one tail) = _____

How many times would you expect T,T if the experiment were repeated

4. 81 times?   5. 162 times?

6. 810 times?   7. 100 times?

8. 200 times?   9. 50 times?

10. 800 times?

PROBABILITY
Problem Set 22

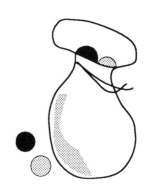

From a bag containing 5 green and 2 red marbles,
one is drawn and then replaced. A second marble is drawn.

1. Draw the two-stage, four-branch tree. List all outcomes
    and their probabilities as fractions and percents.

2. What outcome is most likely?

3. What outcome is least likely?

4. P(the marbles are the same color) = –?–

5. P(the marbles are different colors) = –?–

How many times would you expect two reds if the experiment were repeated

6. 49 times?   7. 98 times?   8. 490 times?   9. 100 times?

10. 50 times?   11. 400 times?   12. 250 times?

Write as decimals.

13. 37%   14. 40%   15. 4%   16. 134%   17. 46.9%   18. 1.8%

For problems 19 and 20, three darts are thrown at the dartboards. Draw the three-stage, eight-branch trees, list the outcomes and determine the probabilities as fractions and whole percents. (Assume the pentagon can be divided evenly into regions half the size of region A.)

19. 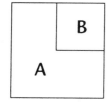   20.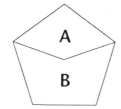

A survey was taken of 345 people concerning their favorite laxative.

21. Two hundred preferred Ze-Lax. What percent is that?

22. Metamuscle was preferred by 73. What percent is that?

23. None preferred Kaster Oil. What percent is that?

GUINNESS RECORD®: The longest run is one of 11,134 miles by Sarah Covington-Fulcher. She began in Laguna Hills, California and ran the perimeter of the United States ending in Los Angeles in 458 days.

24. To the nearest mile, how far did she run each day?

25. How many laps around a 1/4 mile track (similar to those around football fields) did she run each day?

PROBABILITY
Lesson Plan 23

QUIZ   None

OBJECTIVES   Students will be able to mentally calculate 0%, 50%, and 100% of a number. Students will be able to solve alpha-numerics.

MATERIALS NEEDED   1. Transparency 23A and 23B
                        2. Problem Set 23

CLASS ACTIVITIES   One would think that an explanation of 0%, 50%, and 100% would not be necessary. However, this is not the case. A percent line is helpful. Alpha-numerics are introduced on Transparency 23B. Each letter must be replaced by a digit to make a true equation. Repeated letters must be replaced by the same digit.

ASSIGNMENT   Problem Set 23

PROBABILITY 23A

0%, 50%, and 100% of a number

100% of 58 =

50% of 58 =

50% of 62 =

0% of 78 =

100% of 687 =

50% of 640 =

0% of 500 =

PROBABILITY 23B

Crack The Code

$$
\begin{array}{r}
AAA \\
+B \\
\hline
BCCC
\end{array}
$$

$$
\begin{array}{r}
MA \\
+A \\
\hline
AM
\end{array}
$$

$$
\begin{array}{r}
PIT \\
+TIP \\
\hline
QQQ
\end{array}
$$

PROBABILITY
Problem Set 23

Without using a calculator, determine the following.

1. 100% of 64
2. 50% of 64
3. 50% of 28
4. 0% of 78
5. 100% of 564
6. 50% of 7

Write as decimals.

7. 45%
8. 42.8%
9. 9%
10. .6%
11. 174%
12. 100%

The asking price for the tadpole was $654. However, by shrewd bargaining, Winthrop was able to save $68 on the asking price.

13. How much did Winthrop pay for the tadpole?
14. What percent of the asking price did Winthrop pay?
15. What percent of the asking price did Winthrop save?

Of 654 people surveyed, 134 liked Laffer's candy bars.

16. What percent liked Laffer's?
17. What percent did not like Laffer's?
18. How many people did not like Laffer's?

The probability of drawing a red card was 4/11. How many times should Astoria expect to draw a red card if she repeated the experiment

19. 11 times?
20. 22 times?
21. 99 times?

The probability that a slice of bread will fall buttered side down is 82%. How many slices should Slippery Fingers expect to fall buttered side down if...

22. 100 were dropped?
23. 50 were dropped?
24. 800 were dropped?
25. 950 were dropped? *Hint: Look at problems 22–24.*

Keenan tosses a weighted coin
[P(T) = 3/5] three times.

26. Draw the three-stage, eight-branch tree. List all outcomes and calculate their probabilities as fractions and whole percents.
27. What outcome is most likely?
28. P(2H and 1T in any order)
29. P(the coins match)

PROBABILITY
Problem Set 23

How many times should Keenan expect all tails if he does the experiment

30. 125 times?   31. 250 times?   32. 100 times?   33. 400 times?

CRACK THE CODE:

34.   GG
     GG   M is bigger than 6
  +GG
  ――――
   MM

35.   AL
     AL
  + AL
  ――――
  HAL

PROBABILITY
Lesson Plan 24

QUIZ   Quiz 9

OBJECTIVES   Students should be able to determine 10% of a number mentally.

MATERIALS NEEDED   1. Quiz 9
                        2. Transparency 24A and 24B
                        3. Problem Set 24

CLASS ACTIVITIES   Transparency 24A gives examples involving finding 10% of a
                        number while reviewing 0%, 50%, and 100%. Emphasize that these
                        are not calculator problems. 10% = 1/10. This means divide by 10,
                        which tells the student to slide the decimal point so that the
                        number becomes smaller. Telling the students to slide the decimal
                        point to the left is not always meaningful.

ASSIGNMENT   Problem Set 24

PROBABILITY QUIZ 9

Name _____

Two darts are thrown
at the square dartboard.

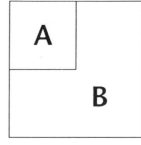

1. Draw the two-stage, four-branch tree writing
    probabilities on the branches. List all outcomes
    and their probabilities as fractions.

2. What outcome is most likely? _____

3. P(both darts land in same area) = _____

4. How many times would you expect both darts
    to land in A if the experiment were repeated

    a. 16 times? _____   b. 32 times? _____

PROBABILITY 24A

0%, 10%, 50%, and 100%

50% of 46

0% of 46

100% of 46

10% of 560

10% of 48

10% of 46.8

50% of 68

100% of 467

PROBABILITY 24B

Gosling surveyed 650 people concerning how they had voted in previous elections.

Draw a percent line to work the following.

0%   100%

1. 100% had voted for Raygun.

    How many is that? _____

2. 10% had voted for Noxin.

    How many is that? _____

3. 50% had voted for Doocockeye.

    How many is that? _____

4. 0% had voted for Mandole.

    How many is that? _____

PROBABILITY
Problem Set 24

For problems 1–4, draw a percent line to answer the question. Do the arithmetic in your head.

1. Fifty percent of the 460 people surveyed owned a VCR. How many owned a VCR?

2. Norval got 10% of the questions correct on the 60 question test. How many questions did Norval get correct?

3. Howell weighed 338 pounds. He lost approximately 10% of his weight. About how many pounds did he lose?

4. Harriet owned 48 harmonicas but sold 50% of them. How many did she sell?

Write as decimals.

5. 56%   6. 60%   7. 6%   8. 45.7%   9. 134%   10. 3.4%

Of 467 people surveyed, 165 preferred Orville Blackenbacher's Popcorn.

11. What percent preferred Orville's?   12. What percent did not prefer Orville's?
13. How many did not prefer Orville's?

The probability that the traffic signal is red is 64%. How many times would Tamara expect it to be red if she approached it

14. 100 times?   15. 50 times?   16. 800 times?

A bag contains seven red and three blue marbles.

17. One is drawn at random. P(R) = –?–

18. One is drawn and replaced, and then a second is drawn. Draw the two-stage, four-branch tree listing all outcomes with their probabilities as fractions and percents.

19. What outcome is most likely?   20. P(marbles are the same color) = –?–

21. P(marbles are different colors) = –?–

How many times should Rhonwen expect two red marbles if the experiment were repeated

22. 100 times?   23. 50 times?   24. 300 times?   25. 1000 times?

CRACK THE CODE:

26.   GO
      GO
     +G
     ─────
     OG

27.   ONE
    + ONE
    ─────
     TWO

BRAIN BUSTER: What number has the same value as the number of letters in its name?

PROBABILITY
Lesson Plan 25

QUIZ   None

OBJECTIVES   Students will use a percent line to mentally calculate 25% of a
                        number. Students will construct a tree diagram with branches of
                        unequal length.

MATERIALS NEEDED   1. Transparency 25A and 25B
                        2. Problem Set 25

CLASS ACTIVITIES   Transparency 25A introduces finding 25% of a number. This is not a
                        calculator procedure. Rather, it involves dividing by four, which is
                        simply done by dividing by two twice.

                        Transparency 25B is an example of a procedure in which the
                        branches on the tree are of unequal length. Once the balloon is
                        broken, the experiment is over. The completed tree should look
                        like this:

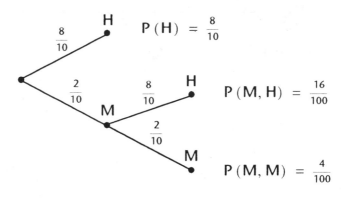

                        Appropriate questions are:
                        P(M,H) = –?–
                        P(M,M) = –?–
                        P(the balloon is broken) = –?–

ASSIGNMENT   Problem Set 25

PROBABILITY 25A

0%, 10%, 25%, 50%, and 100%

Complete the following percent lines.

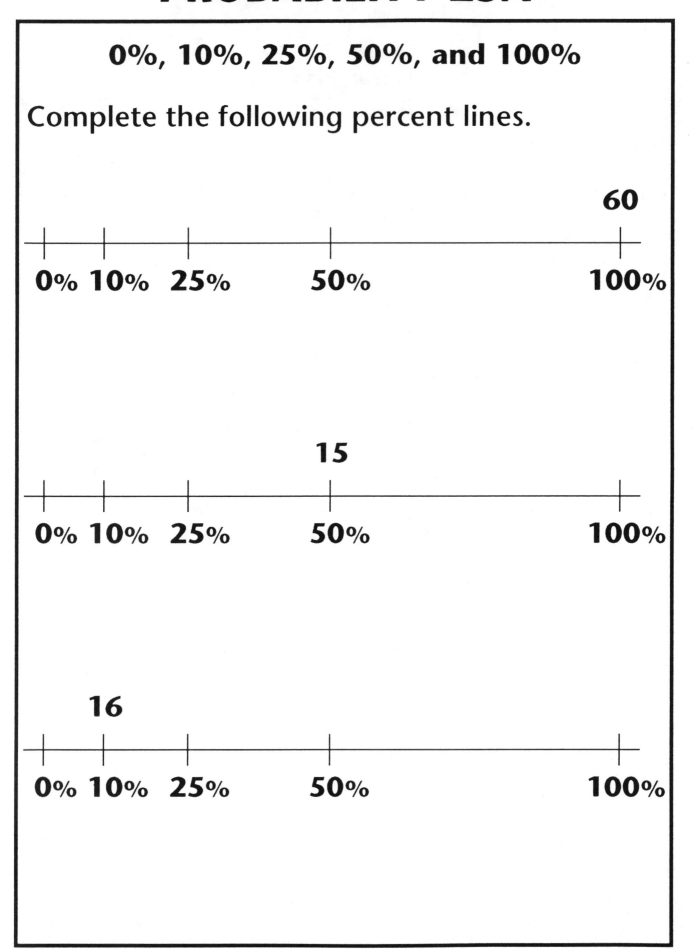

60

| 0% | 10% | 25% | 50% | 100% |

15

| 0% | 10% | 25% | 50% | 100% |

16

| 0% | 10% | 25% | 50% | 100% |

PROBABILITY 25B

Basil and The Balloon

Basil is given two attempts to break a balloon with a dart. His probability of breaking the balloon on any throw is 8/10. What is the probability that Basil will break the balloon?

PROBABILITY
Problem Set 25

For problems 1–3, draw a percent line to answer the question. Do the arithmetic in your head.

In a survey of 480 people concerning fast food preferences:

1. 10% preferred Boogie Queen. How many is that?
2. 50% preferred Quby's. How many is that?
3. 25% preferred Harbee's. How many is that?

In a survey of 465 MacDoogal's customers, only 113 liked their apple pie.

4. What percent liked the apple pie? 5. What percent did not like the apple pie?
6. How many did not like the apple Pie?

CRACK THE CODE:

7.   ICAN
   + ICAN
   ‾‾‾‾‾‾
     DOIT   (all less than 7)

8.   LOOK
    x A
   ‾‾‾‾‾‾
    10101   (K is less than A)

The probability of drawing a three from a deck of cards is 8%. How many times should Alexi expect to draw a three if she repeats the experiment

9. 100 times?   10. 25 times?   11. 500 times?   12. 525 times?

Maldywn throws three darts at the Square Dartboard.

13. Draw the three-stage, eight-branch tree writing the probabilities on the branches. List all outcomes and their probabilities as fractions and percents.

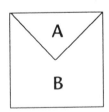

Write the probabilities as percents.

14. P(A,A,A)   15. P(all A's or all B's)   16. P(exactly one A)

How many times should Maldywn expect all darts to land in A if she repeats the experiment

17. 100 times?   18. 500 times?   19. 64 times?   20. 128 times?

PROBABILITY
Problem Set 25

The Aardvarks are playing the Baboons in a best of three games series. The probability of the Aardvarks winning any game is 4/5.

21. Draw a three-stage tree listing all possible outcomes and their probabilities as fractions and percents.

Write the following probabilities as percents.

22. P(Aardvarks win in 2 games)

23. P(Aardvarks win in 3 games)

GUINNESS RECORD®: Jas Angelo (Great Britain) juggled three objects without dropping any for 32,251 seconds in 1989.

24. To the nearest minute, how long did he juggle the objects?

25. Approximately how many hours did he juggle the objects?

PROBABILITY
Lesson Plan 26

QUIZ

Quiz 10

OBJECTIVES

Students use a table of random numbers to simulate an experiment.

MATERIALS NEEDED

1. Quiz 10
2. Transparencies 26A and 26B
3. Experiment 5 (transparency)
4. Table of Random Numbers (transparency)

CLASS ACTIVITIES

Draw a tree exhibiting the three game series on the board. The tree should look like this:

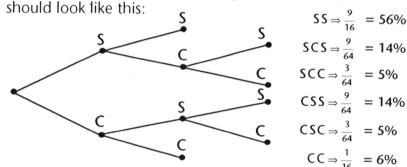

$SS \Rightarrow \frac{9}{16} = 56\%$

$SCS \Rightarrow \frac{9}{64} = 14\%$

$SCC \Rightarrow \frac{3}{64} = 5\%$

$CSS \Rightarrow \frac{9}{64} = 14\%$

$CSC \Rightarrow \frac{3}{64} = 5\%$

$CC \Rightarrow \frac{1}{16} = 6\%$

The probability of the Sparrows winning one game is 3/4. In a three-game series the probability of the Sparrows winning the series is (9/16 + 9/64 + 9/64) 84.375%. Discuss the Five-Game Playoff transparency. Note that the probability of the Sparrows winning the five-game playoff is now 90%. Conclusion: The longer the series, the greater the probability that the stronger team will win.

Experiment 5 simulates a three-game series. To use the random number table, randomly choose a starting point. Then proceed across the page. For example, 48, 05, 64, 89, 47, 42... For this example, 48, 05 means two Sparrow wins and the series is over (S/2). Then 64, 89, 47 means S, C, S and the series is over with a win for the Sparrows (S/3). The instructions on Experiment 5 tell how to determine whether a number represents a win for the Sparrows or Crows. Students should work in pairs. One should work with the random number table and the other should record and determine who the winner is. Each pair of students will simulate 20 series. In the Outcomes section, write the probabilities as percents. The theoretical probabilities come from the first transparency.

ASSIGNMENT

None

PROBABILITY QUIZ 10

Name _____

The probability that a tossed thumbtack will land point up is 78%. How many thumbtacks should Trefor expect to land point up if:

1. 100 are tossed? _____ 2. 50 are tossed? _____

3. 300 are tossed? _____

The probability of a pair of dice totaling 12 is 1/36. How many times should Rhys expect a 12 if the dice are rolled:

4. 36 times? _____   5. 72 times? _____

6. 360 times? _____

Use a percent line in answering the following.

7. Anabel saved 10% on a pumpkin that cost $80.00. How much did Anabel save?

8. Of 346 mathematicians surveyed, 50% preferred Mowton Dont soft drink. How many preferred Mowton Dont?

PROBABILITY 26A

The Playoff

To determine the champion, the Sparrows play the Crows in a best-of-three series. The probability of the Sparrows winning a game is 3/4.

Determine the following probabilities:

P(S win in two games)

P(S win in three games)

P(S win the series)

P(C win in two games)

P(C win in 3 games)

P(C win the series)

PROBABILITY 26B

The Playoff/Five Games

The Sparrows play the Crows in a best-of-five series. The probability of the Sparrows winning any game is 3/4.

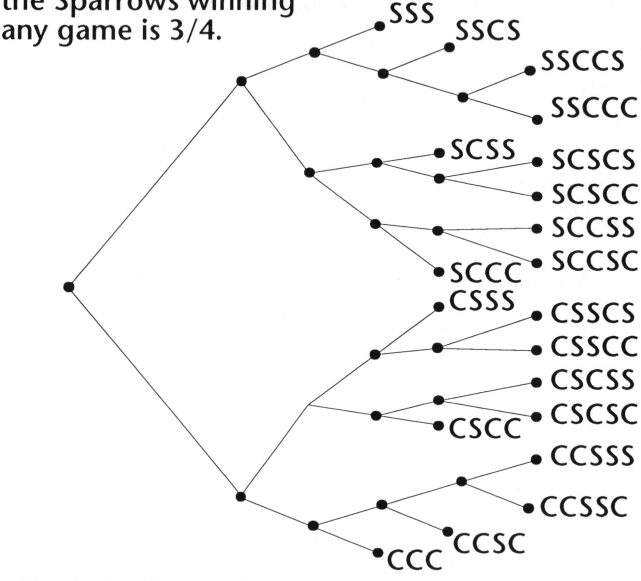

P(S win in 3) = .429

P(S win in 4) = .316

P(S win in 5) = .158

P(S win series) = .903

EXPERIMENT 5

Simulating a Three Game Series

The probability of the Sparrows beating the Crows in one game is 3/4. Use a table of random numbers to simulate a three-game series. Record a win for the Sparrows if you choose a number between 00 and 74 inclusive. A number between 75 and 99 indicates a win for the Crows.

	Digits	Winners	Outcome
1.			
2.			
3.			
4.			
5.			
6.			
7.			
8.			
9.			
10.			
11.			
12.			
13.			
14.			
15.			
16.			
17.			
18.			
19.			
20.			

Outcomes

Pair	Class	Theory
S/2		56%
S/3		28%
C/2		6%
C/3		10%

PROBABILITY
Random Number Table

27 89 70 31 42	52 03 35 60 36	82 05 17 29 06	98 22 45 48 75
59 93 94 48 05	64 89 47 42 96	24 80 52 40 37	20 63 61 04 02
08 42 26 89 53	19 64 50 93 03	23 20 90 25 60	15 95 33 47 64
99 01 90 25 29	09 37 67 07 15	38 31 13 11 65	88 67 67 43 97
12 80 79 99 70	80 15 73 61 47	54 03 23 66 53	98 95 11 68 77
66 06 57 47 17	34 07 27 68 50	36 69 73 61 70	65 81 33 98 85
31 06 01 08 05	45 57 18 24 06	35 30 34 26 14	86 79 90 74 39
85 26 97 76 02	02 05 16 56 92	68 66 57 48 18	73 05 38 52 47
16 78 74 80 93	83 40 59 75 27	66 65 52 22 52	59 60 23 29 49
32 83 36 86 75	48 59 24 05 07	00 45 28 60 37	75 72 76 01 55
52 33 26 64 01	72 06 57 09 61	46 26 87 73 47	43 53 30 17 59
80 55 95 90 68	36 92 21 91 98	96 39 58 47 11	69 14 62 78 26
30 73 21 62 88	08 78 73 95 16	05 92 91 22 30	49 03 14 72 87
30 41 49 11 28	08 56 09 06 53	63 64 39 70 95	38 92 81 24 52
00 58 46 79 93	93 38 18 85 32	23 70 21 17 59	16 49 44 19 38
83 87 83 76 16	08 73 43 25 38	41 45 60 83 32	59 83 01 24 14
71 26 80 95 10	04 06 96 38 27	07 74 20 15 12	33 87 25 01 62
11 71 39 64 16	94 57 91 33 92	25 02 92 61 38	97 19 11 94 75
05 04 98 88 46	62 09 06 83 05	36 56 14 66 35	63 46 71 43 00
57 07 77 51 30	38 20 86 83 92	99 01 68 41 48	27 74 51 80 81
55 07 94 55 99	36 04 98 62 67	93 15 21 04 38	92 41 47 02 06
18 83 39 37 57	80 43 07 35 21	38 95 35 43 53	77 53 19 82 05
06 74 20 52 97	19 14 63 80 17	96 59 12 90 08	18 49 05 13 42
05 26 62 91 29	14 17 72 98 49	89 59 43 00 95	41 80 11 95 06
82 83 85 09 80	06 59 07 52 63	27 52 46 56 88	63 87 34 55 79

PROBABILITY
Lesson Plan 27

QUIZ   None

OBJECTIVES   Students will be able to mentally determine 5% and 15% of a
                    number. Students will be able to differentiate between situations
                    that lead to equal and unequal length branch trees.

MATERIALS NEEDED   1. Transparencies 27A and 27B
                    2. Problem Set 27

CLASS ACTIVITIES   Transparency 27A leads the students through a procedure that
                    allows them to mentally calculate 5% of a number and 15% of
                    a number. To calculate 5% of a number, first calculate 10% and
                    then take half of the result. To determine 15% of a number, add 5%
                    and 10%.

                    The two-shot situation in basketball leads to a tree with equal length
                    branches. The one-and-one situation leads to a tree with unequal
                    branches.

ASSIGNMENT   Problem Set 27

PROBABILITY 27A

5% and 15%

Determine the following:

100% of 60

50% of 60

25% of 60

10% of 60

5% of 60

15% of 60

10% of 40

5% of 40

15% of 40

Sales tax in Indiana is 5%. A fair tip at a restaurant is 15%.

Determine the sales tax on a $80.00 pair of tennis shoes.

The bill for Carly and Daryl's dinner was $18.00. How much tip should they leave?

PROBABILITY 27B

Fraser The Free Throw Shooter

In the big game, Fraser is fouled and awarded two free throws. The probability that Fraser will make a free throw is 4/5. Determine the following probabilities:

P(he will score 2 points)

P(he will score 1 point)

P(he will not score)

Later in the game he is fouled and is awarded one-and-one. Determine the following probabilities:

P(he will score 2 points)

P(he will score 1 point)

P(he will not score)

PROBABILITY
Problem Set 27

What is the amount of sales tax (5% in Indiana) that Newman would have to pay on items 1–6:

1. $60.00 sweat shirt
2. $5.00 calculator
3. $460.00 CD player
4. $8950.00 car
5. $36,780 yacht
6. 89¢ pencil

Vanslow interviewed 480 Teen Age Mutant Ninja Turtles.

7. 50% watched "WKRP in Cleveland." How many watched the show?
8. 10% watched "Triplet Peaks." How many is that?
9. 193 watched "Rescue 123." What percent is that?
10. 78% watched "Late Night with David Numberman." What percent didn't watch the show?

Driving to work, Penny encounters two traffic lights. The probability that a light is green is 5/8.

11. Draw the two-stage, four-branch tree listing all outcomes as fractions and percents.

Write the following probabilities as percents.

12. P(Penny does not have to stop)
13. P(Penny stops exactly once)
14. P(Penny has to stop twice)

How many times should Penny expect to stop twice if she drives to work

15. 100 times?
16. 50 times?
17. 400 times?
18. 450 times?

The probability that Dead Eye Drina will make a free throw is 90%.

19. Write 90% as a simplified fraction.
20. What is the probability that Drina will miss a free throw?
21. Drina is awarded two free throws. Draw the two-stage tree, listing all outcomes and their probabilities as fractions and whole percents.
22. Drina is awarded one-and-one. Draw the tree with unequal length branches, listing all outcomes and their probabilities as fractions and whole percents.
23. *Guess and Test:* Marle had nickels and quarters in her pocket totaling $1.90. She had eight more nickels than quarters. How many of each coin did she have?
24. *Make a List:* Elram had nickels and quarters in his pocket totaling $1.90. List the seven combinations of coins that Elram might have had?

CRACK THE CODE:

25.   LOVLVEL
          x L
       ─────────
       87654321

PROBABILITY
Lesson Plan 28

QUIZ   Quiz 11

OBJECTIVES   Students will be able to draw a tree diagram for a two-stage
                        experiment in which the two trials are dependent.

MATERIALS NEEDED   1. Quiz 11
                        2. Transparency 28
                        3. Problem Set 28

CLASS ACTIVITIES   Transparency 28 gives examples that show the difference between
                        replacement and non-replacement. Draw the trees for all examples.
                        Notice that the denominators in the second stage branches are all
                        one less than those in the first stage branches. The tree for
                        Experiment 2 is shown below.

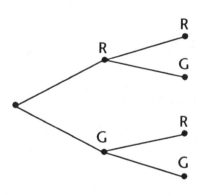

RR	$\frac{20}{42}$
RG	$\frac{10}{42}$
GR	$\frac{10}{42}$
GG	$\frac{2}{42}$

ASSIGNMENT   Problem Set 28

PROBABILITY QUIZ 11

Name _____

The probability of Manley being in school on any day is 86%. How many days would you expect Manley to be in school if school has been in session:

1. 100 days? _____   2. 50 days? _____

The probability of a box of Cookie Jacks containing a prize is 3/65. How many prizes should Adler expect to win if he buys

3. 65 boxes? _____   4. 130 boxes? _____

Of 500 people interviewed:

5. 25% watched "Weeks of Our Lives." How many is that? _____

6. 50 watched "The Old and the Restless." What percent is that? _____

7. The probability of Twyla making a free-throw is 9/10. Draw a tree for a one-and-one situation. List all outcomes and probabilities.

PROBABILITY 28

Replacement Versus Nonreplacement

A bag contains five red and two green marbles.

Experiment 1:

1. Choose a marble.

2. Replace the marble.

3. Choose a second marble.

Experiment 2:

Same as Experiment 1, except don't replace the first marble.

A bag contains six red and one green marble.

Experiment 3:

Repeat Experiment 1 with this new bag.

Experiment 4:

Repeat Experiment 2 with this new bag.

PROBABILITY
Problem Set 28

For problems 1–6, draw a two-stage tree. List outcomes and probabilities as fractions and percents.

1. A deck of cards consists of four cards—two Aces and two Kings. A card is drawn and replaced. A second card is drawn.

2. The same as problem 1, except the first card is not replaced before a second is drawn.

3. A deck of cards consists of three cards—two Aces and a King. A card is drawn and replaced. A second card is drawn.

4. The same as problem 3, except the first card is not replaced before a second is drawn.

5. A bag contains 12 red and 8 green marbles. A marble is drawn and replaced. A second marble is drawn.

6. The same as problem 5, except the first marble is not replaced before a second is drawn.

Determine the 5% sales tax that would be charged on the following items.

7. A $800.00 stereo   8. A $6856.00 ring   9. A $6.98 earring   10. A 47¢ goldfish

Of 568 Martians interviewed, 25% watch "The Tonight Show with Johnny Autoson."

11. How many watch the show?   12. What percent don't watch the show?

13. How many do not watch "The Tonight Show?"

Of 4734 Venusians interviewed, 3456 watch "Tire of Fortune."

14. How many do not watch the show?   15. What percent watch the show?

16. What percent do not watch the show?

Determine a fair tip for the following bills at Deryn's Diner.

17. $10.00   18. $28.00   19. $68.00   20. $130.00

Write as decimals.

21. 45%   22. 70%   23. 7%   24. 5.9%

GUINNESS RECORD®: The world's greatest bird watcher or "twitcher" is Harvey Gilson of Lausanne, Switzerland who had logged 6,713 of the 9,016 known species by April 10, 1990.

25. What percent of the known species had he logged?

26. How many species had he not logged?

BRAIN BUSTER: Draw a straight line on the clock, so that the numbers on one side of the line have the same sum as those on the other side.

PROBABILITY
Lesson Plan 29

QUIZ   Quiz 12

OBJECTIVES   The student will be able to draw a tree showing outcomes in a
                        three-stage experiment where non-replacement is involved.

MATERIALS NEEDED   1. Quiz 12
                        2. Transparency 29
                        3. Problem Set 29

CLASS ACTIVITIES   The example on Transparency 29 is a three-stage, non-replacement
                        problem. Emphasize that the denominators continue to decrease at
                        each stage of the experiment.

ASSIGNMENT   Problem Set 29

PROBABILITY QUIZ 12

Name _____

Determine each of the following:

1. 25% of 80 = _____

2. 10% of 70 = _____

3. 10% of $64.00 = _____

4. 5% of $86.00 = _____

5. 5% of 89¢ = _____

6. 15% of $30.00 = _____

In a survey of 660 car owners, 25% preferred Firerock tires.

7. How many preferred Firerock? _____

8. What percent did not prefer Firerock? _____

9. How many did not prefer Firerock? _____

PROBABILITY 29

Non-Replacement—Three Draws

A deck of cards consists of four Aces and three Kings.

A A A A K K K

Three cards are drawn *without* replacement.

Determine the following probabilities.

P(all Aces) = _____

P(all Kings) = _____

P(the cards match) = _____

P(two Aces and one King) = _____

PROBABILITY
Problem Set 29

A deck of cards has four Aces, four Kings, and two Jokers. Thor draws two cards without replacement.

1. Draw a two-stage, nine-branch tree. List outcomes with probabilities as fractions and percents.

Write the following probabilities as percents.

2. P(two Aces)
3. P(two Jokers)
4. P(cards match)
5. P(cards don't match)
6. P(at least one Joker)

How many times would Thor expect to draw two Jokers if he repeated the experiment

7. 100 times?
8. 300 times?
9. 50 times?
10. 450 times?

A bag contains seven red and three green marbles. Schuyler draws three marbles without replacement.

11. Draw a three-stage, eight-branch tree. List outcomes with probabilities as fractions and percents.

Write the following probabilities as percents.

12. P(all red)
13. P(all green)
14. P(two green and one red)
15. P(all same color)
16. P(not same color)

In a survey of 4600 marathoners, 25% wore Sadida running shoes.

17. How many wore Sadida shoes?
18. What percent didn't wear Sadida shoes?
19. How many did not wear Sadida running shoes?

In a survey of 850 luge riders, 376 wore Ekin shoes.

20. What percent wore Ekins?
21. How many did not wear Ekins?
22. What percent did not wear Ekins?

Clotilda bought a new pair of Ekins for $130.00.

23. Determine the 5% sales tax.
24. What was Clotilda's total bill?
25. Clotilda paid for the shoes with seven $20 bills. How much change will she get?

The bill for the brunch at Eppie's Eatery was $18.00.

26. How much should Sybel leave for a tip?
27. Sybel has $20.00. Can she leave a fair tip?

CRACK THE CODE:

28. PAT
    + PTA
    ‾‾‾‾‾‾
     TAP   (P = 4)

29. PPQ
     x Q
    ‾‾‾‾‾‾
    AQBQ   (A, B, P, and Q are consecutive)

PROBABILITY
Lesson Plan 30

QUIZ   None

OBJECTIVES   Review in preparation for Test 3.

MATERIALS NEEDED   1. Transparency 30
                    2. Problem Set 30

CLASS ACTIVITIES   In Transparency 30 students can think about the effect of adding the green socks. It turns out that when the distribution is even, it is difficult to get a match. If the distribution were heavily skewed, say 98 of one color and two of the other color, it would be easy to get a match. The answer to the Bonus is three. This leads into a discussion of the Birthday Problem. Although there need to be 367 people in a room in order to guarantee two matching birthdays, with 23 people the probability is greater than 50%, and with 30 people the probability is more than 90%. (Elementary probability books explain this phenomena.) Check for duplicate birthdays in your class.

ASSIGNMENT   Problem Set 30

PROBABILITY 30

Sorcha and Her Socks

Sorcha has 8 red and 4 green socks in a drawer. In the morning, it is dark in his room and she blindly chooses one and then a second. Determine the probability that she chooses a matched pair.

To improve her probability of choosing a matched pair, Sorcha buys two more pair of green socks. What is her probability of choosing a pair now?

BONUS!!! How many socks would Sorcha need to pull out of the drawer to guarantee getting a matched pair?

PROBABILITY
Problem Set 30

1. Mandrake the Magician has a hat with five black and two white rabbits in it. He pulls out a rabbit, replaces it and pulls out a second rabbit. Draw the two-stage, four-branch tree and list all outcomes and their probabilities as fractions and percents.

2. This time Mandrake does not replace the first rabbit before pulling out the second. Draw the two-stage, four-branch tree and list all outcomes and their probabilities as fractions and percents.

3. At the carnival, Robyn is given two chances to toss a ring over a peg. The probability that she will do this on any try is 5/6. If she does it on her first try, she does not get a second chance. Draw the two-stage, three-branch tree and list all outcomes and their probabilities as fractions and percents.

If five cards are drawn from a standard deck, the probability that they will contain a pair is 42%. How many times should Pascale expect to get a pair if he repeats the experiment

4. 100 times? 5. 50 times? 6. 300 times? 7. 450 times?

If a letter is chosen at random from the alphabet, the probability that it is a vowel is 5/26. How many times should Sterling expect to chose a vowel if he repeats the experiment

8. 26 times? 9. 260 times? 10. 13 times?

Determine the following.

11. 100% of 64   12. 50% of 6   13. 0% of 56   14. 10% of 540
15. 5% of 540   16. 15% of 540 17. 25% of 84   18. 50% of 22

Alton got 65 out of 83 correct on the BIG math test.

19. What percent did she get correct?   20. How many did she miss?
21. What percent did she miss?

Bastian got 10% of the 40 questions correct on the big math test.

22. How many questions did Bastian get correct?
23. How many did Bastian miss?   24. What percent did Bastian miss?

CRACK THE CODE:

25. SS
    x 4
    ―――
    ASK

PROBABILITY
Test 3

Do not write on this test. Do all work on scratch paper and put the answers in the appropriate space on the answer sheet.

Two darts are thrown at the dartboard.

1. Draw the tree, and list outcomes with probabilities as fractions and whole percents.
2. What outcome is most likely?
3. What is the probability that the second dart will land in the same region as the first dart?

A bag contains two red and four green marbles. Arabella draws one, replaces it, and draws a second.

4. Draw the tree, and list outcomes with probabilities as fractions and whole percents.
5. What outcome is most likely?
6. What is the probability that the marbles match?

How many times should Arabella expect two red marbles if she repeats the experiment

7. 100 times?   8. 50 times?   9. 700 times?   10. 350 times?

A bag contains two red and three green marbles. Hudson draws one, and without replacing it, draws a second.

11. Draw the tree and list outcomes with probabilities as fractions.
12. What is the probability that the marbles are different colors?

How many times should Hudson expect to draw two red marbles if he repeats the experiment

13. 20 times?   14. 10 times?   15. 70 times?

Basketball Babette is given three attempts to make a basket at a carnival. She shoots until she makes a basket or misses all three. The probability of her making a basket on any try is 2/3.

16. Draw the tree (unequal length branches) and list all outcomes with their probabilities as fractions and whole percents.

Answer the following in percents.

17. P(miss all three)   18. P(she makes a basket)

Write as whole percents.

19. 63/100   20. 11/25   21. 11/37

22. In working 19–21, you should have used your calculator only once. Which problem was a calculator problem?

PROBABILITY
Test 3

If five cards are drawn from a standard deck the probability of getting two pair is about 5%. How many times should Keisha expect two pair if she repeats the experiment

23. 100 times?   24. 400 times?   25. 20 times?

The probability of rolling a total of three with a pair of dice is 1/18. How many times should Mervyn expect a total of three if he rolls the dice

26. 18 times?   27. 180 times?   28. 36 times?

During the basketball season, Malvina tried 79 field goals and made 53.

29. What percent did she make?   30. What percent did she miss?
31. How many did she miss?

A group of 540 prison inmates were surveyed concerning their TV viewing habits.

32. 10% watched Top Cops. How many watched Top Cops?
33. 50% watched Unsolved Mysteries. How many watched it?
34. 25% watched FBI: Untold Stories. How many watched it?
35. 100% watched Rescue 911. How many watched it?

On the big Math test, Jessy missed 10% of the 40 questions.

36. How many questions did she miss?   37. How many questions did she get right?
38. What percent of the questions did she get right?

Determine the amount of sales tax (5%) charged on

39. an $80.00 Corvette.   40. a $5682 hamburger.
41. Determine the tip (15%) that Handel should leave for a meal costing $40.

CRACK THE CODE:

42.   RAT
    + TAR
    ‾‾‾‾‾
      EEEK

43.   SCAR
        x R
    ‾‾‾‾‾
      RACS

PROBABILITY

Test 3 Answer Sheet

Name _____

1.

•

2. _____

3. _____

4.

•

5. _____

6. _____

7. _____

8. _____

9. _____

10. _____

11.

•

12. _____

13. _____

14. _____

15. _____

16.

•

17. _____   32. _____

18. _____   33. _____

19. _____   34. _____

20. _____   35. _____

21. _____   36. _____

22. _____   37. _____

23. _____   38. _____

24. _____   39. _____

25. _____   40. _____

26. _____   41. _____

27. _____   42.

28. _____

29. _____   43.

30. _____

31. _____

PROBABILITY
Lesson Plan 31

QUIZ   None

OBJECTIVES   Students will be able to determine the probability of specific events in two-stage experiments without drawing a tree. Students will be able to work percent problems of the type "*a* had characteristic 1, *b* had characteristic 2. What percent of the total had characteristic 1?"

MATERIALS NEEDED   1. Transparencies 31A and 31B
                        2. Problem Set 31

CLASS ACTIVITIES   In this unit, we will not be focusing on all outcomes of an experiment, but only specific outcomes. Therefore, it is not necessary to draw a complete tree. Transparency 31A gives three examples. The first experiment is a review from Unit One.

                        The second experiment introduces the new idea. Answers are as follows:

                        P(even, even) = 3/7 x 3/7 = 9/49
                        P(even, odd) = 3/7 x 4/7 = 12/49
                        P(3,odd) = 1/7 x 4/7 = 4/49
                        P(3,3) = 1/7 x 1/7 = 1/49

                        In the third experiment, the first disk is not replaced. Hence, the denominator in the second fraction is diminished by one and so is the numerator (sometimes). The answers to the four examples are 6/42, 12/42, 3/42, and 0/42.

                        Transparency 31B gives examples of percent problems slightly different than those previously encountered.

ASSIGNMENT   Problem Set 31

PROBABILITY 31A

The Derby and The Disks

A hat has seven disks in
it numbered from 1–7.

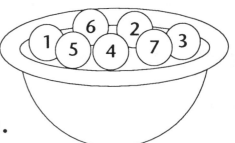

Experiment 1: A disk is chosen.

P(even) =

P(more than 5) =

P(even *and* more than 5) =

P(even *or* more than 5) =

Experiment 2: A disk is chosen, replaced and a
second disk is chosen.

P(even,even) =   P(even,odd) =

P(3,odd) =   P(3,3) =

Experiment 3: A disk is chosen, not replaced,
and a second disk is chosen.

P(even,even) =   P(even,odd) =

P(3,odd) =   P(3,3) =

PROBABILITY 31B

Percent

On the quiz, Jehu got 56 correct and 25 wrong. What percent did he get correct?

In a survey of students concerning their favorite TV detective, 346 students chose Ruglock while only 87 preferred Joe Monday. What percent chose Ruglock?

PROBABILITY
Problem Set 31

For problems 1–20, a bowl contains disks numbered from 1–15.
Write all of the probabilities as fractions.

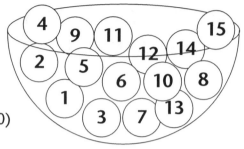

Experiment 1: One disk is drawn.

1. P(even)
2. P(4)
3. P(greater than 15)
4. P(last digit is 3)
5. P(even, more than 10)
6. P(even or more than 10)
7. P(multiple of 6)
8. P(less than 20)

Experiment 2: A disk is drawn, replaced and a second disk is drawn.

9.   P(even,even)
10. P(even,7)
11. P(less than 3,2)
12. P(4,even)
13. P(4,odd)
14. P(4,4)

Experiment 3: A disk is drawn, not replaced and a second disk is drawn.

15. P(even,even)
16. P(7,even)
17. P(7,odd)
18. P(5,5)
19. P(3,less than 5)
20. P(even, odd)

The probability that Percy Pupil will be in school on any day is 9/10. It is a short week with only two days of school. Answer 22–24 in percents.

21. Draw a two-stage, four-branch tree listing all possible outcomes and their probabilities as fractions and percents for the two-day period.
22. P(in school both days)   23. P(misses both days)   24. P(misses one day)

Lilac missed seven questions and got 43 correct.

25. What percent did Lilac get correct?

In an interview of high school students, 17 students had seen "Back to the Past," and five had not.

26. What percent had seen it?   27. What percent had not seen it?

GUINNESS RECORD®: In 24 hours Jeff Liles made 17,227 free throws and missed 7,082 at Castleberry High School, Fort Worth, TX.

28. How many free throws did he shoot?   29. How many did he shoot each hour?
30. How many did he shoot each minute?   31. What percent did he make?

PROBABILITY
Lesson Plan 32

QUIZ

None

OBJECTIVES

Given a sequence of numbers, the student will be able to determine the pattern and find the next three terms in the sequence

MATERIALS NEEDED

1. Transparencies 32A and 32B
2. Problem Set 32

CLASS ACTIVITIES

Transparency 32A reviews ideas from the previous lesson.

Transparency 32B asks students to determine the next term in a sequence. The rules for the examples are:

add 2
add 3
add 3, add 4, add 5,...
add 1, add 2, add 1, add 2,...
subtract 2, add 8, subtract 2, add 8...
1 one, 2 twos, 3 threes...
subtract 6, subtract 5, subtract 4...

ASSIGNMENT

Problem Set 32

PROBABILITY 32A

The Big Draw

A standard deck consists of 52 cards. There are 13 cards 2, 3, 4,…, K, A in each of 4 suits ♥, ♦, ♣, and ♠.

1. Greeta draws two cards, replacing the first before she draw the second.

    P(red,red) =

    P(A♦,red) =

    P(A,A) =

    P(2,face card) =

2. Magnolia draws two cards but does not replace the first before drawing the second.

    P(red,red) =

    P(A♦,red) =

    P(A,A) =

    P(2,face card) =

PROBABILITY 32B

What's Next?

1, 3, 5, 7…

2, 5, 8, 11…

2, 5, 9, 14…

4, 5, 7, 8, 10, 11…

3, 1, 9, 7, 15, 13…

1, 2, 2, 3, 3, 3, 4…

26, 20, 15, 11, 8…

PROBABILITY
Problem Set 32

Five cards from 1–5 are in a bowl. Alkta chooses one.
Write the following probabilities as simplified fractions.

1. P(odd)   2. P(less than 8)
3. P(even and greater than 3)
4. P(even or greater than 3)

Five cards from 1–5 are in a bowl. Helmuth chooses one, replaces it, and chooses a second.

5. P(odd,odd)   6. P(2,5)   7. P(greater than 4,less than 3)
8. P(3,3)

Five cards from 1–5 are in a bowl. Pat chooses one, then without replacing it, chooses a second.

9. P(odd,odd)   10. P(2,5)   11. P(greater than 4,less than 3)
12. P(3,3)   13. P(3,odd)   14. P(3,even)

Write the next three terms in the pattern.

15. 4, 9, 14, 19…   16. 1, 3, 9, 27…   17. 8, 11, 15, 20, 26…
18. 56, 50, 44, 38…   19. 64, 32, 16, 8…

Althena got 42 votes and Galena got 58 votes.

20. How many people voted?   21. What percent of votes did Althena receive?
22. What percent of the votes did Galena receive?

Quintella encountered 17 stoplights on her drive to school. Twelve of them were red.

23. What percent were red?   24. What percent were green?

Algernon draws two cards from a standard deck without replacement. Determine the following probabilities as percents.

25. P(red,red)   26. P(♦,♦)

How many times should Algernon expect to draw two red cards if he repeats the experiment

27. 100 times?   28. 50 times?   29. 300 times?   30. 80 times?

BRAIN BUSTER: Al spent twice as much as Beth. Candi spent $3 more than Dael. Beth spent $4 less than Candi. Dael spent $42. How much did Al spend?

PROBABILITY
Lesson Plan 33

QUIZ   Quiz 13

OBJECTIVES   Given a decimal, students will be able to write it as percent rounded
                        to the nearest tenth.

MATERIALS NEEDED   1. Quiz 13
                        2. Transparencies 33A and 33B
                        3. Problem Set 33

CLASS ACTIVITIES   Use Transparency 33A to show the difference between rounding to
                        a whole percent and to a tenth of a percent. Note that the correct
                        answer to the last example is 15.0% not 15%.

                        In working through Transparency 33B, it is necessary to round to
                        the nearest tenth. If not, the answer to the last problem is 0%.

ASSIGNMENT   Problem Set 33

PROBABILITY QUIZ 13

Name _____

A bag contains slips numbered from 1–5. Two are drawn without replacement. Write the following as whole percents.

1. P(odd,odd) = _____

2. P(3,even) = _____

3. P(3,odd) = _____

4. P(3,3) = _____

On the biology test, Helmut got 35 correct and 15 wrong.

5. How many questions were on the test? _____

6. What percent did Helmut get correct? _____

7. What percent did Helmut get wrong? _____

PROBABILITY 33A

Rounding Percents to Tenths

Decimal	Percent	
	Whole	Tenth
.57463		
.986443		
.59742		
.37428		
.15983		

PROBABILITY 33B

Euchre Deck

Spades ♠	9 10 J Q K A
Hearts ♥	9 10 J Q K A
Clubs ♣	9 10 J Q K A
Diamonds ♦	9 10 J Q K A

Two cards are drawn in succession.

P(red,red) =

P(♥,♥) =

P(♥,red) =

P(A,A) =

P(A♦,A) =

PROBABILITY
Problem Set 33

Write the next three terms in the sequence.

1. 1, 1, 1/2, 2, 1/3, 3, 1/4...   2. 3, 11, 19, 27...   3. 2, 6, 18, 54...
4. 5, 4, 10, 9, 15, 14, 20...   5. 1, 22, 333, 4444...   6. 80, 40, 20, 10...

Determine each of the following.

7. 25% of 44   8. 10% of 60   9. 5% of 60   10. 15% of 60
11. 50% of 74   12. 100% of 43

Delhonde draws one card from a 24-card Euchre deck (cards nine through Ace). Write the following probabilities as fractions and whole percents.

13. P(A)   14. P(face card)   15. P(♠)   16. P(red)
17. P(red and an A)   18. P(red or an A)

Write as percents rounded to the nearest tenth.

19. .34719   20. .8626   21. .0007   22. .749684327

Kalphe draws two cards from a Euchre deck without replacement. Write the following probabilities as fractions and percents rounded to the nearest tenth.

23. P(A,A)   24. P(♦,♦)   25. P(red,red)   26. P(9,10)
27. P(A♦,♦)   28. P(A♦,black)

In a survey of 65 people, 47 preferred AquaStale while the rest preferred Coolgrate.

29. What percent preferred AquaStale?   30. What percent preferred Coolgrate?
31. How many preferred Coolgrate?

Fifty elves preferred Ting Breakfast drink and 31 preferred Moonkist oranges.

32. What percent preferred Ting?   33. What percent preferred Moonkist oranges?

GUINNESS RECORD®: The strongest animals in the world are the larger beetles of the Scarabaeidae, found mostly in the tropics. One of these, the rhinoceros beetle, can support 850 times its own weight. On the other hand, an elephant can support 25% of its own weight.

34. If a 140-pound teenager were as strong as this beetle, how much could she support?

35. If a 140-pound teenager were as strong as an elephant, how much could he support?

PROBABILITY
Lesson Plan 34

QUIZ   None

OBJECTIVES   Given a sequence of instructions, students will be able to determine
                        what black hole a number will disappear into.

MATERIALS NEEDED   1. Transparency 34
                        2. Problem Set 34

CLASS ACTIVITIES   Transparency 34 describes what a scientist means by a black hole. In
                        our problems a black hole is a value that every number disappears
                        into. For each example, have each student choose a number and go
                        through the procedures listed in the examples. In Example 1, every
                        number should disappear into the black hole 4.

                        The black hole for Example 2 is 6. It is instructive for students to use
                        some easy fractions as their starting numbers.

ASSIGNMENT   Problem Set 35

PROBABILITY 34

A Black Hole

A black hole is a region in space so dense, and with gravity so intense, that nothing, not even light, can escape. If an object gets near it, it will always disappear into it. Some mathematical procedures lead to black holes.

Procedure 1: Choose a number. Multiply by 6, add 12, divide by 3, and subtract twice the original.

Procedure 2: Choose a number. Multiply by 4, add 12, divide by 2, and subtract twice the original.

PROBABILITY
Problem Set 34

● Write the next three terms in the sequence.

1. 1, 4, 9, 16...

2. 1, 3, 4, 6, 7, 8, 10, 11...

3. 5, 10, 15, 20, 25...

4. a, b, d, e, g...

5. a, z, b, y, c...

6. a, c, b, d, f, e, g...

Write as percents rounded to the nearest tenth.

7. .76543

8. .649674

9. .1306

10. .0049

11. .128333

12. 1.87549834

The procedures below lead to black holes. Find them by going through each procedure twice.

13. Choose a number.
    Multiply by 4.
    Add 8.
    Divide by 2.
    Subtract twice the original.
    Black hole= ?

14. Choose a number.
    Add 5.
    Multiply by 4.
    Subtract 20.
    Divide by the original.
    Black hole = ?

● Hanley draws three cards from a deck without replacement. Write the following probabilities as fractions and percents rounded to the nearest tenth.

15. P(red,red,red)

16. P(♦,red,red)

17. P(♦,♦,red)

18. P(♦,♦,♦)

Gylthe made $470 last year. He received a 10% raise.

19. How much was the raise?

20. How much is he making this year?

21. Kenza makes $5.00 and hour. However, 25% of her salary goes to taxes, social security, and insurance. What is her net hourly salary?

A jar contains eight red and two white marbles. Randel draws two without replacement. Write the following probabilities as fractions and percents rounded to the nearest tenth.

22. P(red,red)

23. P(white,white)

24. P(red,white)

25. P(white,red)

A survey was taken of 467 athletes. 356 drank Crockade.

26. What percent drank Crockade?

27. What percent did not drink Crockade?

28. How many did not drink Crockade?

● *BRAIN BUSTER:* Move three coins so that the arrow points down.

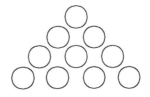

PROBABILITY
Lesson Plan 35

QUIZ   Quiz 14

OBJECTIVES   Students will develop Pascal's Triangle and then will be able to apply it in probability situations.

MATERIALS NEEDED   1. Quiz 14
                        2. Transparencies 35A and 35B
                        3. Problem Set 35

CLASS ACTIVITIES   There are two ways to introduce Pascal's Triangle. One way is to simply start revealing it one row at a time and ask the students to discover the elements in the next row by looking for some of the many patterns that exist.

     1. rows are symmetric (build up and down)
     2. first and last elements are 1
     3. second element is row number
     4. any element is sum of the two elements above it
     5. row sums double

The following procedure develops the triangle and shows how it is related to a probability situation: Toss two coins. Examine the outcomes using a tree but focus only on the number of heads.

    $P(2H) = 1/4$
    $P(1H) = 2/4$
    $P(0H) = 1/4$

Do the same thing with three and then four coins. The results are as follows:

    $P(3H) = 1/8$ $P(4H) = 1/16$
    $P(2H) = 3/8$ $P(3H) = 4/16$
    $P(1H) = 3/8$ $P(2H) = 6/16$
    $P(0H) = 1/8$ $P(1H) = 4/16$
                    $P(0H) = 1/16$

Note that the numerators are the elements of the triangle and the denominators are the row sums.

Again, let the students guess the big rows. Pascal's Triangle can be used to determine probabilities in experiments in which there are a number of independent repeated trials with two outcomes, each having a probability of 1/2.

(continued on next page)

PROBABILITY
Lesson Plan 35

To use Pascal's Triangle in working the first example, write out the eighth row of the triangle with the associated events written above/below the numbers.

1	8	28	56	70	56	28	8	1	256
8H	7H	6H	5H	4H	3H	2H	1H	0H	

In working the second example, write out the sixth row of the triangle.

1	6	15	20	15	6	1	64
6B	5B	4B	3B	2B	1B	0B	

ASSIGNMENT   Problem Set 35

PROBABILITY QUIZ 14

Name_____

Write the next three terms.

1. 4, 9, 14, 19, _____, _____, _____

2. 11, 21, 20, 30, 29, 39, _____, _____, _____

3. 4, 3/4, 5, 3/5, 6, 1/2, 7, 3/7, _____, _____, _____

4. a, c, e, g, i, _____, _____, _____

Write as percents rounded to tenths.

5. .73448 = _____

6. .283676 = _____

7. .3896 = _____

8. Thirty-seven voted for Barton and 13 voted against him. What percent voted for him?

    _____

Petrina drew two cards in succession from a 24 card euchre deck (no replacement). Write as fractions.

9. P(red,red) = _____   10. P(A♦,♣) = _____

PROBABILITY 35A

Pascal's Triangle

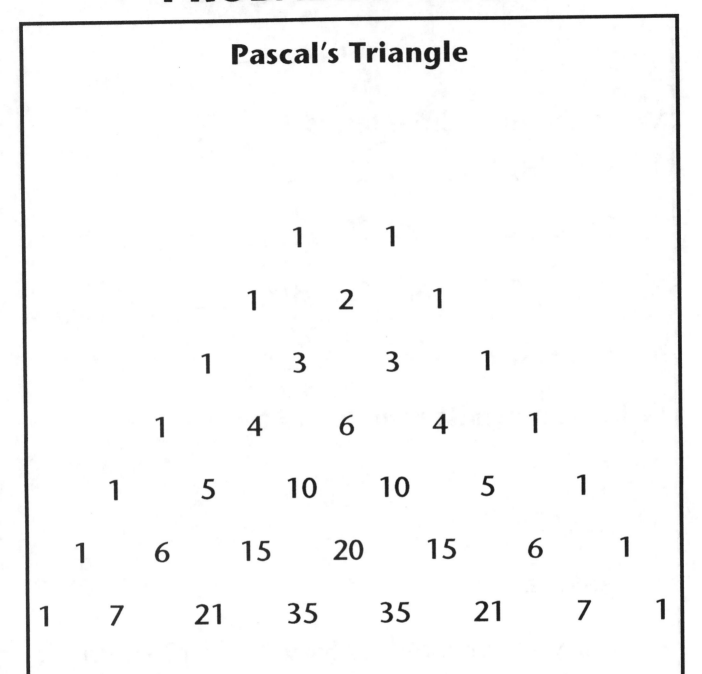

PROBABILITY 35B

Applications of Pascal's Triangle

Eight coins are tossed.

P(8T) =

P(5T) =

P(3H) =

P(more than 5T) =

A family consists of 6 children.

P(6 boys) =

P(4 boys) =

P(2 girls) =

P(more girls than boys) =

PROBABILITY
Problem Set 35

Each of the following procedures leads to a black hole. Determine the black hole by going through the procedure twice.

1. Choose a number.
    Add 6.
    Multiply by 3.
    Subtract 18.
    Divide by the original.
    Black hole = ?

2. Choose a number.
    Multiply by itself.
    Subtract 4.
    Divide by 2 less than the original.
    Subtract the original.
    Black hole = ?

Write the next three terms in the sequence.

3. 1, 0, 2, 0, 3, 0, 4...

4. a, b, z, c, d, y, e...

5. 1, a, 3, c, 5, e, 7...

6. 1/8, 1/4, 3/8, 1/2, 5/8...

A deck consists of six cards: the J, Q, and K of spades and clubs. McGirlver draws two without replacement. Write the following probabilities as percents rounded to the nearest tenth.

7. P(K,K)

8. P(K♠,♠)

9. P(♠,♠)

How many times should McGirlver expect two spades if the experiment were repeated

10. 100 times?

11. 50 times?

12. 400 times?

Determine the following.

13. 5% of $48.00

14. 15% of $48.00

15. 25% of $48.00

16. 50% of $48.00

Enlith tosses a fair coin five times.

17. Write out the fifth row of Pascal's Triangle.

18. P(5H)

19. P(3H)

20. P(2T)

21. P(5T)

22. P(more than 2H)

23. P(5H or 5T)

Ichabod guesses at seven questions on a true-false test.

24. Write the seventh row of Pascal's Triangle.

25. P(7 wrong)

26. P(5 correct)

27. P(2 wrong)

28. P(6 or 7 correct)

29. P(more than 50% correct).

GUINNESS RECORD®: The most prolonged continuous shower bath by a woman is 121 hours and 1 minute by Lisa D'Amato of Harpur College, Binghamton, New York beginning on November 5, 1981.

30. How many days did she shower?

31. If she used up a bar of soap every seven hours, about how many bars did she use up?

PROBABILITY
Lesson Plan 36

QUIZ

None

OBJECTIVES

Students will perform Experiment 6 and compare the experimental results with the theoretical results obtained by using Pascal's Triangle.

MATERIALS NEEDED

1. Transparency 36
2. Experiment 6. (Transparency also.)
3. A styrofoam cup and five coins for each pair of students.

CLASS ACTIVITIES

Have each student work through the instructions on Transparency 36 on a sheet of paper (possibly the back of Experiment 6) with you. If they do the work correctly, the six digit number that they end up with will be their birth date.

In experiment 6, each pair of students should shake the five coins in the cup and then pour them on to the desk. Record the number of heads that show. Do this for approximately 10 minutes. Next, summarize the class totals and calculate the experimental probabilities and compare them with the theoretical probabilities obtained from the fifth row of Pascal's Triangle. Write the probabilities obtained from the Triangle both as fractions and percents.

ASSIGNMENT

None

PROBABILITY 36

Birthday Black Hole

1. Write the number of your birth month. _____

2. Multiply by 4. _____

3. Add "Unlucky" 13. _____

4. Multiply by 25. _____

5. Subtract 200. _____

6. Add the day of the month you were born on. _____

7. Double. _____

8. Subtract 40. _____

9. Multiply by 50. _____

10. Add the last two digits of your birth year. _____

11. Subtract a number that is between 10,499 and 10,501. _____

EXPERIMENT 6

The Five Coin Toss

The purpose of this experiment is to compare the theoretical probabilities with the experimental probabilities when five coins are tossed.

Outcome	Tally (PAIR)	PAIR Total	CLASS Total	%	THEORY
5H					
4H					
3H					
2H					
1H					
0H					

To determine the theoretical probabilities we will use the fifth row of Pascal's Triangle.

1 5 10 10 5 1 Sum: 32

PROBABILITY
Lesson Plan 37

QUIZ   None

OBJECTIVES   Students will be able to use a calculator to determine the percent of
                        a number.

MATERIALS NEEDED   1. Transparency 37
                        2. Problem Set 37

CLASS ACTIVITIES   I suggest the following procedure. For each series of problems, draw
                        a percent line and enter the answers to the mental problems on the
                        line. Estimate the answers to the calculator problems using the
                        percent line. Use the calculator to obtain the exact answer.

                            17% of 80 =

                            .17 x 80 =

ASSIGNMENT   Problem Set 37

PROBABILITY 37A

Calculating Percents

10% of 80 =

25% of 80 =

17% of 80 =

8% of 80 =

50% of 880 =

100% of 880 =

73% of 880 =

92% of 880 =

50% of 64 =

25% of 64 =

10% of 64 =

37% of 64 =

8% of 64 =

19% of 64 =

PROBABILITY
Problem Set 37

In problems 1–4, do parts a and b mentally and then use a calculator in working part c. The answer to part c should be between the answers to part a and b. Write the answers to the nearest whole number.

1a. 10% of 60	2a. 50% of 48	3a. 10% of 40	4a. 50% of 38
b. 25% of 60	b. 25% of 48	b. 15% of 40	b. 100% of 38
c. 17% of 60	c. 35% of 48	c. 13% of 40	c. 67% of 38

Og spins the spinner seven times.

5. Write out the seventh row of Pascal's Triangle and label each number in terms of Red. (For example, 1R, 2R, etc.)

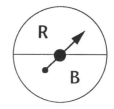

Write the probabilities as fractions and whole percents.

6. P(7R) 7. P(0R) 8. P(3R or 4R) 9. P(2B)
10. P(same number of R as B)

How many times should Og expect five Red if he conducts the experiment

11. 100 times? 12. 700 times?

Disks numbered 1–6 are placed in a fish bowl. Swaley draws two without replacement. Write the following probabilities as whole percents.

13. P(4,even) 14. P(4,odd) 15. P(4,4)

Mahala draws two of the disks but replaces the first before drawing the second. Write the following probabilities as whole percents.

16. P(4,even) 17. P(4,odd) 18. P(4,4)

Write the next three terms of the sequence.

19. 1, 5, 9, 13, 17... 20. a, 2, b, 4, c, 6, d... 21. a, b, 3, d, e, 6, g...
22. 2, 10, 50, 250... 23. 1/10, 1/5, 3/10, 2/5, 1/2...

In a survey of 4678 inmates, 3456 watched "Some in the Family."

24. What percent watched "Some in the Family"?
25. What percent did not watch the show?
26. How many did not watch the show?

PROBABILITY
Lesson Plan 38

QUIZ   Quiz 15

OBJECTIVES   Students will use the calculator appropriately in solving percent
                        problems of the form "p% of P is."

MATERIALS NEEDED   1. Quiz 15
                        2. Transparencies 38A and B
                        3. Problem Set 38

CLASS ACTIVITIES   Transparency 38A is a fun problem used to get class started.
                        Transparency 38B gives examples of the new problems. The
                        television shows are soap operas. Emphasize that the calculator
                        should not be used in the first two problems in the first example.
                        A percent line should be drawn so that the answers to the first two
                        can be used to determine if the answers to the last two problems
                        are reasonable. The first two problems in the second example
                        should be done mentally.

ASSIGNMENT   Problem Set 38

PROBABILITY QUIZ 15

Name _____

Complete the sixth row of Pascal's Triangle.

1. 1 6 15 20 ___ ___ ___

Complete the eighth and ninth rows.

2. 1 8 ___ ___ 70 56 ___ ___ 1

3. ___ ___ 36 ___ ___ ___ ___ ___ ___ ___ ___

The fourth row of Pascal's Triangle is:

   1 4 6 4 1   <u>16</u>

A family consists of four children. Write the probabilities as fractions.

4. P(4 boys) = _____

5. P(4 girls) = _____

6. P(2 or 3 boys) = _____

7. P(more boys than girls) = _____

8. The most likely family will consist of _____ girls and _____ boys.

PROBABILITY 38A

While exploring the backwoods, Sarita discovered a cone shaped object two feet tall. (Enter 2) To Sarita's misfortune, it was the home of 2669 angry insects. (Enter x 2669) Who were the insects? (Turn the calculator upside down!!!)

What city in Idaho has a population of 35108?

PROBABILITY 38B

Television Viewing

In a survey of 4594 wrestlers:

    10% watched "Old and Content."
    50% watched "Two Lives to Live."
    37% watched "Timid and Ugly."
    65% watched "As the World Rotates."

Determine the number that watched each show.

The probability that a person chosen at random watches "Sargent Hospital" is 13%. How many people would you expect to watch the show if you surveyed:

    100 people?

    300 people?

    456 people?

    2876 people?

PROBABILITY
Problem Set 38

Tasmine goes through five stoplights on her way to work. The probability that any light is green is 1/2.

1. Write out the fifth row of Pascal's Triangle and label each number in terms of green lights.

Write the following as fractions and whole percents.

2. P(5G)   3. P(3G)   4. P(2R)   5. P(more than 2 green)

How many times should Tasmine expect to drive to work without stopping (5G) if she makes the trip

6. 100 times?   7. 600 times?   8. 50 times?   9. 438 times?

Determine the following percentages. Write the answers as whole numbers. You should use your calculator three times.

10. 50% of 86   11. 25% of 86   12. 37% of 86   13. 10% of 57
14. 100% of 67   15. 41% of 89   16. 7% of 78   17. 50% of 842

Artina draws two cards without replacement from a deck of 20 cards composed of the A, 2, 3, 4, and 5 of each suit. Determine the probabilities as percents rounded to the nearest tenth.

18. P(red,red)   19. P(♦,♦)   20. P(♦,red)   21. P(A,A)   22. P(A♦,A)

How many times should Artina expect to draw two Aces if she repeats the experiment

23. 100 times?   24. 500 times?   25. 467 times?

26. An airplane flying at 33,000 feet (enter 33000) encounters engine trouble and starts falling at the rate of 2821 feet per minute (subtract 2821). How did the pilot arrive safely at the airport? Turn your calculator upside down!

GUINNESS RECORD®: The greatest documented number of schools attended by a pupil is 265, by Wilma Williams from 1933–1943 when her parents were in show business in the United States.

27. On the average, how many schools did she attend each year during this ten-year period?

28. School is in session for about 180 days each year. About how many days was she in each school during a year?

BRAIN BUSTER: A leak in the roof allows one drop through the first night, two the second night, four the third night, eight the fourth night, 16 the fifth night, etc. When will the 300th drop fall through?

PROBABILITY
Lesson Plan 39

QUIZ

None

OBJECTIVES

Students will discover patterns that involve Pascal's Triangle.

MATERIALS NEEDED

1. Transparencies 39A and 39B
2. Problem Set 39

CLASS ACTIVITIES

The problem is to determine how many ways words are spelled when you must start at the top and always move down.

To count the number of ways that *Wow* is spelled, we look at the bottom row of W's. There is one way to end at the first W, two ways to end at the middle W, and one way to end at the last W. Hence, there are four paths that spell *Wow*. A similar process with *Neat* leads to one, three, three, and one or a total of eight ways. *Super* has one, four, six, four, and one path giving a total of 16 ways. To answer the question for *Awesome*, we observe that *Neat* has four letters and we use the third row of Pascal's Triangle. Also, the sum of the third row is 2^3. Hence the number of paths for *Awesome* is the sum of the sixth row of the triangle which is $2^6 = 64$ ways.

Super...has 34 letters. Hence, the total number of paths is 2^{34} which is more than 8.5 billion. You might spend some time discussing how big a billion is. Ask how long it would take to count to a billion, or a million.

ASSIGNMENT

Problem Set 39

PROBABILITY 39A

Pascal Patterns

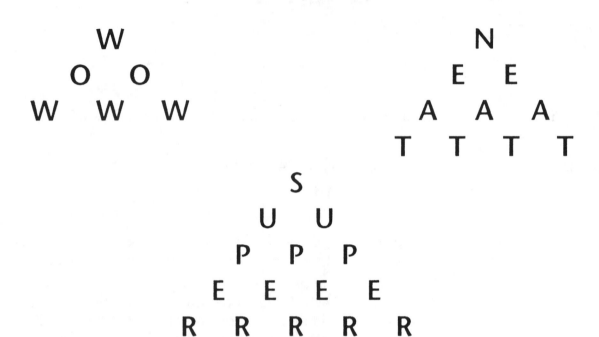

How about *awesome?*

How about *supercalifragilisticexpealadocious.*

PROBABILITY 39B

```
                  S
                 U U
                P P P
               E E E E
              R R R R R
             C C C C C C
            A A A A A A A
           L L L L L L L L
          I I I I I I I I I
         F F F F F F F F F F
        R R R R R R R R R R R
       A A A A A A A A A A A A
      G G G G G G G G G G G G G
     I I I I I I I I I I I I I I
    L L L L L L L L L L L L L L L
   I I I I I I I I I I I I I I I I
  S S S S S S S S S S S S S S S S S
 T T T T T T T T T T T T T T T T T T
I I I I I I I I I I I I I I I I I I I
C C C C C C C C C C C C C C C C C C C C
E E E E E E E E E E E E E E E E E E E E E
X X X X X X X X X X X X X X X X X X X X X X
P P P P P P P P P P P P P P P P P P P P P P P
E E E E E E E E E E E E E E E E E E E E E E E E
A A A A A A A A A A A A A A A A A A A A A A A A A
L L L L L L L L L L L L L L L L L L L L L L L L L L
A A A A A A A A A A A A A A A A A A A A A A A A A A A
D D D D D D D D D D D D D D D D D D D D D D D D D D D D
O O O O O O O O O O O O O O O O O O O O O O O O O O O O O
C C C C C C C C C C C C C C C C C C C C C C C C C C C C C C
I I I I I I I I I I I I I I I I I I I I I I I I I I I I I I I
O O O O O O O O O O O O O O O O O O O O O O O O O O O O O O O O
U U U U U U U U U U U U U U U U U U U U U U U U U U U U U U U U U
S S S S S S S S S S S S S S S S S S S S S S S S S S S S S S S S S S
```

PROBABILITY
Problem Set 39

Four hundred sixty English professors were surveyed.

1. Exactly 50% had read *Tale of Three Towns*. How many had read it?
2. Exactly 10% had read *Moby Richard*. How many had read it?
3. About 37% had read *Call of the Tame*. How many had read it?
4. One hundred fifteen had read *A Thanksgiving Carol*. What percent had read it?
5. Only 31 had read *Uncle Fred's Mansion*. What percent had read it?

Write the next three terms in the sequence.

6. 1, 2, 4, 7, 11, 16...   7. 1, 2, 4, 8, 16, 32...   8. b, a, d, c, f, e...

A hat contains six slips of paper labeled A, B, C, D, E, and F. Dunkling draws two without replacement. Write each of the probabilities as a fraction and whole percent.

9. P(vowel,vowel)   10. P(A,vowel)   11. P(A,A)   12. P(A,B)

13. P(vowel,consonant)

14. How many ways is Math spelled?   15. How many ways is Spell spelled?

```
           M                                        S
          A A                                      P P
         T T T                                    E E E
        H H H H                                  L L L L
                                               L L L L L
```

The probability that Norna will visit his grandmother on any Sunday is 13%. About how many times would you expect Norna to visit her in

16. 100 Sundays?   17. A year (52 Sundays)?   18. A decade (10 years)?

19. A score (20 years)?

Delmar the Daschund had a litter of five puppies. Assume that the probability that a puppy is male is 1/2.

20. Write the fifth row of Pascal's Triangle. Label each number in terms of male puppies.

Write each of the probabilities as fractions and percents rounded to the nearest tenth.

21. P(5M)   22. P(all same sex)   23. P(2F or 3F)   24. P(more F than M)

25. Last week 18,900 people (enter 18900) paid $20.00 (Enter x 20) to watch a movie made in 1919 (Enter + 1919). What did this film make them do?

PROBABILITY
Lesson Plan 40

QUIZ

None

OBJECTIVES

Students will be able to discover patterns that involve Pascal's Triangle.

MATERIALS NEEDED

1. Transparencies 40A and 40B
2. Problem Set 40

CLASS ACTIVITIES

The rules for spelling the words are as follows:

You must start on the outside and move toward the center and down. If you write down the number of ways to spell the word above the starting letter, Pascal's Triangle will appear.

There are 7 ways.

The answer for GOAT is 15 ways.

The answer for ZEBRA is 31 ways.

ASSIGNMENT

Problem Set 40

PROBABILITY 40A

Patterns With Pascal

How many ways is "Cow" spelled?

```
        C
      C O C
    C O W O C
```

How many ways is "Goat" spelled?

```
          G
        G O G
      G O A O G
    G O A T A O G
```

How many ways is "Zebra" spelled?

```
          Z
        Z E Z
      Z E B E Z
    Z E B R B E Z
  Z E B R A R B E Z
```

PROBABILITY 40B

```
                    E
                 E  L  E
              E  L  E  L  E
           E  L  E  P  E  L  E
        E  L  E  P  H  P  E  L  E
     E  L  E  P  H  A  H  P  E  L  E
  E  L  E  P  H  A  N  A  H  P  E  L  E
E  L  E  P  H  A  N  T  N  A  H  P  E  L  E
```

PROBABILITY
Problem Set 40

1. Colville scored 99.9% on his drivers test (Enter 99.9). He attributed his score to two factors (Enter ÷ 2). One was his IQ of 135 (Enter ÷ 135) and the other was his Zodiac sign. What was his sign?

2. How many ways is Cubs spelled?

```
        C
      C U C
    C U B U C
  C U B S B U C
```

3. How many ways is Tigers spelled?

```
          T
        T I T
      T I G I T
    T I G E G I T
  T I G E R E G I T
T I G E R S R E G I T
```

4. If Yankees were written in a similar triangular pattern, how many ways would it be spelled?

Write the next three terms in the sequences.

5. 7/11, 8/13, 3/5, 10/17, 11/19…

6. 1a, 4d, 7g, 10j, 13m…

Preston tossed four coins.

7. Write the fourth row of Pascal's Triangle and label each number in terms of the number of heads that show.

Write each of the following as fractions and whole percents.

8. P(4H)

9. P(3H)

10. P(all T or all H)

11. P(equal number of H and T)

How many times should Preston expect 4 heads if he repeated the experiment

12. 100 times?

13. 50 times?

14. 348 times?

A bowl contains three red, two white, and five blue marbles. Vanetta draws one, replaces it, and draws a second. Write each of the following as fractions and whole percents.

15. P(red,red)

16. P(red,white)

17. P(white,not white)

18. P(not blue,not blue)

Using the same bowl as Vanetta, Marmaduke chooses two marbles without replacement. Write each of the following as fractions and percents rounded to the nearest tenth.

19. P(red,red)

20. P(red,white)

21. P(white,not white)

22. P(not blue,not blue)

PROBABILITY
Problem Set 40

Ronna surveyed 700 mathematics teachers.

23. 50% read Newsmonth. How many is that?
24. 25% read Sports Pictures. How many is that?
25. 46% read Persons magazine. How many is that?
26. 87% read Rolling Pebble. How many is that?

PROBABILITY 40A

Patterns With Pascal

How many ways is "Cow" spelled?

```
        C
      C O C
    C O W O C
```

How many ways is "Goat" spelled?

```
        G
      G O G
    G O A O G
  G O A T A O G
```

How many ways is "Zebra" spelled?

```
          Z
        Z E Z
      Z E B E Z
    Z E B R B E Z
  Z E B R A R B E Z
```

PROBABILITY
Test 4

Do not write on this test. Do all work on scratch paper and put the answer in the appropriate space on the answer sheet.

Write the next two terms in each sequence.

1. 1, 2, 4, 5, 7, 8, 10...   2. 1, 5, 9, 13, 17...   3. 3, 6, 12, 24, 48...

4. z, a, y, b, x, c, w...   5. 4/7, 5/8, 2/3, 7/10, 8/11...

Write each of the following as percents rounded to the nearest tenth.

6. .86321   7. .12368   8. .05555   9. .169797

How many different paths are there spelling the words?

10.   M   11.   M
          M A M   A A
      M A T A M   T T T
    M A T H T A M   H H H H

12. Perdita ran 673 yards (Enter 673) in 58 seconds (Enter x 58) and would have set a world record for the mile run except for an obstacle. What was the obstacle?

For each of the following, choose a number and determine the black hole it disappears into.

13. Choose a number.   14. Choose a number.
    Multiply by 5.   Add 4.
    Add 20.   Multiply by 6.
    Divide by four more than   Subtract 24.
    the original.   Divide by 6.
    Subtract 4.   Subtract the original.
    Black hole = ?   Black hole = ?

15. Write out the fifth row of Pascal's Triangle.

On his way to his job at the deodorant factory, Denzil encounters six traffic signals. Assume that the probability of any light being green is 1/2. Use the sixth row of Pascal's Triangle in answering questions 16–20. Write answers as fractions.

                        1 6 15 20 15 6 1

16. P(all Green)   17. P(4G)   18. P(2R)   19. P(5 or 6 Green)

20. P(same number green as red)

The probability that Tegwin the Teacher chooses a tie that matches his shirt is 12%.

21. What is the probability that his outfit does *not* match?

PROBABILITY
Test 4

How many times should Tegwin expect to get a matching outfit if he goes through this procedure

22. 100 times?   23. 50 times?   24. 700 times?   25. 850 times?
26. 386 times?

Cards numbered from 1–8 are placed in a hat. Wilmer draws a card, replaces it, and draws a second card. Write each of the following as fractions.

27. P(even, even)   28. P(4, even)   29. P(4, 4)   30. P(3, more than 5)
31. P(multiple of 4, 7)

Abdul repeats the same experiment as Wilmer except she does not replace the first card before the second is drawn. Write each of the following as fractions.

32. P(even, even)   33. P(4, even)   34. P(4, 4)   35. P(3, more than 5)
36. P(multiple of 4, 7)

A survey was taken of 600 history teachers.

37. 50% had heard of George Washingpound. How many is that?
38. 10% had heard of Millard Filless. How many is that?
39. 35% had heard of Richard Nixoff. How many is that?
40. 150 had heard of Tom Edidaughter. What percent is that?
41. 578 had heard of Harry Falseman. What percent is that?

The meal cost Minty $22.

42. Determine the 5% tax on the meal.
43. Minty wants to leave a tip that is 15% of the cost of the meal. How much should he leave?

Write the answers to 44–48 as whole percents. Two cards are drawn from a deck without replacement.

44. P(red, red)   45. P(♣, black)   46. P(♣, ♣)
47. Anrinda made 48 shots and missed two. What percent did she make?
48. On the test, Katrina got 456 out of 487 questions correct. What percent did she get correct.

By purchasing a $48 goldfish on sale, Eglof was able to save 25%.

49. How much money did Eglof save?   50. How much did he pay for the goldfish?

PROBABILITY

Test 4 Answer Sheet

Name _____

1. _____

2. _____

3. _____

4. _____

5. _____

6. _____

7. _____

8. _____

9. _____

10. _____

11. _____

12. _____

13. _____

14. _____

15. _____

16. _____

17. _____

18. _____

19. _____

20. _____

21. _____

22. _____

23. _____

24. _____

25. _____

26. _____

27. _____

28. _____

29. _____

30. _____

31. _____

32. _____

33. _____

34. _____

35. _____

36. _____

37. _____

38. _____

39. _____

40. _____

41. _____

42. _____

43. _____

44. _____

45. _____

46. _____

47. _____

48. _____

49. _____

50. _____

PROBABILITY
Lesson Plan 41

QUIZ   Quiz 16

OBJECTIVES   Students will be able to determine the value of a game.

MATERIALS NEEDED   1. Quiz 16
                        2. Experiment 7 (Transparency also)
                        3. A styrofoam cup and three cubes (two of one color and the third of another color) for each pair of students
                        4. Transparency 41
                        5. Problem Set 41

CLASS ACTIVITIES   *These activities take two class periods.*

                        Start class by explaining Experiment 7. The outcome is either same or different, and the winner is either A or B. The last column keeps A's cumulative winnings. We need to see how A stands after 20 games. In my class, we played 200 games and A was a total of $54 in the hole. This means that his average loss (loss per game) was $54 ÷ 200 = $.27. It appears that this is not a fair game. In a fair game one should break even. On the back of Experiment 7, determine the value of the game played in Experiment 7. To determine the value of a game, multiply the probabilities of the outcomes by the payoff and then sum (considering signs) the results.

                        For Experiment 7, the probability of a match is 1/3 and the probability of no match is 2/3. Confirm these results with a tree diagram. The value of the game is:

                        Winnings   1/3 x ($1) =   33¢
                        Losses   2/3 x (⁻$1) =   ⁻67¢
                        Value Of The Game   =   ⁻34¢

(continued on next page)

PROBABILITY
Lesson Plan 41

This means that in the long run, A would expect to lose 34¢ every time the game is played. Compare with experimental results. The games described on the Transparency have values of ⁻50¢, 50¢, and 0¢. The last game is a fair game. Students with good insight can reason in the following manner about the first game. In four games, I will lose three times ($12) and then win once ($10). Therefore, after four games I will be $2 in the hole for an average loss of 50¢. The method outlined for Experiment 7 confirms this intuitive idea. A tree is necessary to determine the probabilities for Game Three.

Quiz 16 should be given the second day.

ASSIGNMENT   Problem Set 41

PROBABILITY QUIZ 16

Name_____

Winfred plays a game in which he is paid $5 for a win, but if he loses he must pay $1. The chart shows the outcomes of 10 games.

GAME	win/lose	A's winnings
1	w	_____
2	l	_____
3	l	_____
4	l	_____
5	w	_____
6	l	_____
7	w	_____
8	l	_____
9	l	_____
10	w	_____

1. Complete the right hand column which keeps track of Winfred's winnings.

2. Does this game seem to favor Winfred? _____

3. What was his average winning per game? _____

4. What percent of games did Winfred win? _____

5. What percent of the games did he lose? _____

PROBABILITY 41

Dartboard Games

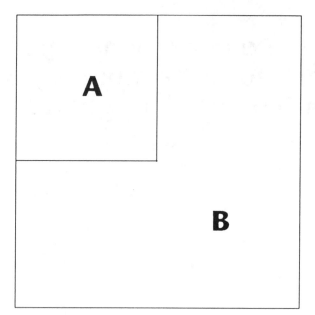

Game 1: Throw a dart.

Lands in A, win $10.

Lands in B, lose $4.

Should you play?

Game 2: Throw a dart.

Lands in A, win $20.

Lands in B, lose $6.

Should you play?

Game 3: Throw two darts.

Both in A, win $72

One in A, break even.

Both in B, lose $8.

Should you play?

EXPERIMENT 7

Colored Cubes

A cup contains two cubes of one color and one of a second color. Two are chosen without replacement. If they match, player A wins $1.00. If they're different, A loses $1.00. It sounds like the game is fair... but is it? Play the game 20 times. Keep track of how much A wins.

Game	Outcome	Winner	A's Winnings
1.	_____	_____	_____
2.	_____	_____	_____
3.	_____	_____	_____
4.	_____	_____	_____
5.	_____	_____	_____
6.	_____	_____	_____
7.	_____	_____	_____
8.	_____	_____	_____
9.	_____	_____	_____
10.	_____	_____	_____
11.	_____	_____	_____
12.	_____	_____	_____
13.	_____	_____	_____
14.	_____	_____	_____
15.	_____	_____	_____
16.	_____	_____	_____
17.	_____	_____	_____
18.	_____	_____	_____
19.	_____	_____	_____
20.	_____	_____	_____

PROBABILITY
Problem Set 41

For problems 1–5 determine the value of the game.

A bowl contains 40 slips of paper, five labeled *Win* and 35 labeled *Lose*. Alfont chooses one at random.

1. Game 1: Draw a *Win* and the prize is a $32 sweater. Draw a *Lose* and you pay $4.
2. Game 2: Draw a *Win* and the prize is $80. Draw a *Lose* and you pay $12.

Ginger spins the spinner once.

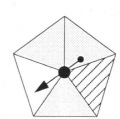

3. Game 1: Land on striped and you win $100.
    Otherwise you lose $20.
4. Game 2: Land on dotted and you win $80.
    Otherwise you lose $100.
5. Game 3: Land on dotted and you win $60.
    Land on plain and you win $100. Land on striped and you lose $300.

The value of a game is $3. How much should Eloph expect to win if he plays the game

6. 10 times?   7. 17 times?   8. 342 times?
9. Adonis plays a game five times and wins a total of $40. What was his average win?
10. Rackin plays a game seven times and loses a total of $28. What was her average loss?

In a large box of pens, 94% will work. How many can Garth expect to work if he chooses a group of

11. 100?   12. 50?   13. 700?   14. 458?   15. 2387?

A box contains 47 good prunes and 36 rotten ones.

16. What percent are rotten?   17. What percent are good?

A box contains 460 widgets. 10% are broken.

18. How many are broken?   19. What percent are not broken?
20. How many are not broken?

GUINNESS RECORD®: Paddy Doyle did 37,350 push-ups in a 24 hour period at the Holiday Inn, Birmingham, United Kingdom on May 1 and 2, 1989.

21. How many is that each hour?   22. How many is that each minute?

PROBABILITY
Lesson Plan 42

QUIZ   None

OBJECTIVES   Students will be able to determine the value of a game when all of
                        the outcomes are positive. Students will be able to complete a three
                        by three magic square.

MATERIALS NEEDED   1. Transparency 42
                        2. Problem Set 42

CLASS ACTIVITIES   The Rutabaga Sale is modeled after sales that actually occur.
                        The value of the game is:

                        1/50 x $50 + 5/50 x $25 + 20/50 x $10 + 24/50 x $5 = $9.90.

                        In a three by three magic square, all rows, columns, and diagonals
                        have the same sum.

ASSIGNMENT   Problem Set 42

PROBABILITY 42

The Rutabaga Sale

At the rutabaga sale, Cresley found out how much she saved by drawing a slip of paper from a jar by the cash register. The jar contained 50 slips of paper with the savings written on them.

Savings	Number
$50	1
$25	5
$10	20
$5	24

How much should Cresley expect to save?

Magic Squares

8	1	6
3	5	7
4	9	2

12	2	16
	10	

PROBABILITY
Problem Set 42

For problems 1, 2, and 3, determine the value of the game. In each game, a pair of tetrahedral dice are rolled. The table shows the possible outcomes.

+	1	2	3	4
1	2	3	4	5
2	3	4	5	6
3	4	5	6	7
4	5	6	7	8

1. If the sum is greater than 5, win $30. Otherwise, lose $10.

2. If the sum is even and more than 6, win $100. Otherwise, lose $7.

3. If the sum is a multiple of 4, win $100. Otherwise, lose $25.

The value of a game is $2. How much should Gustav expect to win if he plays the game

4. 7 times?   5. 20 times?   6. 893 times?

The value of a game is $3.50. How much should Kiana expect to lose if she plays the game

7. 10 times?   8. 67 times?   9. 963 times?

10. Siobahn played a game six times and won $18 dollars. What was her average winning per game?

11. Tamsin played the same game eight times and lost $6. What was his average loss per game?

A survey was taken of 500 doctors.

12. 25% had heard of Pistols 'n Petunias. How many is that?

13. 38% had heard of C. M. Wrench. How many is that?

14. 95% had heard of Strawberry Sleet. How many is that?

15. 149 had heard of Angryonna. What percent is that?

Copy and complete the Magic Squares.

16.

10		
	7	9
		4

17.

4		
1.5		
2		1

18. The winning team averaged 187.904 m.p.h. (Enter 187.904). For their efforts, they divided the first prize of $500 (Enter ÷ 500). What event did they win?

PROBABILITY
Lesson Plan 43

QUIZ

None

OBJECTIVES

Given sufficient information, students will be able to complete a four by four Magic Square.

MATERIALS NEEDED

1. Transparency 43
2. Problem Set 43

CLASS ACTIVITIES

A popular way to raise money is to raffle a house.

The information on the top half of the Transparency came from an article in the Fort Wayne Journal-Gazette, Sunday, November 24, 1991. The value of the game is ⁻$65.36 if the cash prize of $175,000 is taken instead of the $225,000 house. In a four by four Magic Square, there are many arrangements of four numbers that have the same sum: the four corners, the four center, the two pair in the middle of the opposite sides, the four in each corner.

ASSIGNMENT

Problem Set 43

PROBABILITY 43

St. Joseph Medical Center
Auxillary Raffle Payoff

3500 tickets each costing $125...

Twenty Prizes

1    $225,000 home or $175,000 cash

1   1992 car valued at $15,000

3    $2500 cash

5    $1000 cash

5    $500 cash

5    $250 cash

What is the expected value of this game?

Four by Four Magic Squares

16	2	3	13
5	11	10	8
9	7	6	12
4	14	15	1

4		17	3
	9		12
			8
			15

PROBABILITY
Problem Set 43

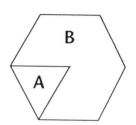

1. Zorah throws a dart at the dart board. If it lands in A, she wins $12. Otherwise, she loses $3. Determine the value of the game.

Richmal throws two darts at the dart board.

2. P(both land in A)   3.   P(both land in B)

4. If they both land in A, she wins $50. If they both land in B, she loses $2. Otherwise, she breaks even. Determine the value of the game.

Sorcha plays a game 20 times. Determine her average win or loss if at the end of the 20 games, she is

5. $40 ahead.   6. $10 behind.   7. $50 behind.   8.   even.

Ranald plays a game 30 times. How much money should he expect to end up with if the value of the game is

9. $3?   10. $1.60?   11. 43¢?

12. There was a big sale at the supermarket yesterday. The item came in packages of 8 (Enter 8) and sold for $4.38 a package (Enter x 4.38). What was the item?

A survey was taken of 300 vegetarians concerning their favorite singer.

13. 194 chose Paula Abdul. What percent is that?
14. 13% chose Mariah Carey. How many is that?
15. None chose Sinead O'Connor. What percent is that?

Copy and complete the Magic Squares.

16.

7		5
2		
3		

17.

4		
$\frac{1}{2}$		
3	$3\frac{1}{2}$	

18.

18	4	5	15
	13	12	
11			
	16		

BRAIN BUSTER: Use 31 coins to make a dollar. Use the same number of nickels as pennies.

PROBABILITY
Lesson Plan 44

QUIZ   Quiz 17

OBJECTIVES   Students will be able to determine the percent increase/decrease.
                        None of the problems involve the use of a calculator.

MATERIALS NEEDED   1. Quiz 17
                        2. Transparency 44
                        3. Problem Set 44

CLASS ACTIVITIES   Percent increase/decrease is determined by dividing the change by
                        the original amount. For the first example, the result is:

                        $(106 - 100)/100 = 6\%$.

                        It should be pointed out that the first magic square in the
                        assignment requires that the students use negative numbers.

ASSIGNMENT   Problem Set 44

PROBABILITY QUIZ 17

Name _____

1. Rawbur plays a game six times and wins a total of $24. What was his average win? _____

2. Uticam plays a game having a value of $^-$$1.50. How much should he expect to lose if he plays the game 40 times? _____

3. Horace throws a dart at the dartboard. If it lands in A, he wins $15. If it lands in B, he loses $6. Determine the value of the game.

A	B

4. Complete the three by three Magic Square.

11		
13	9	
3		

PROBABILITY 44

Percent Increase/Decrease

The number of holsteins increased from 100 to 106. What is the percent increase?

There were 50 guernseys last year, but only 39 this year. Determine the percent decrease.

In 1990 there were 20 jerseys, but in 1991 there were 27. What was the percent increase?

The number of brown swiss increased from 450 to 900. Determine the percent increase.

PROBABILITY
Problem Set 44

For problems 1–6, determine the percent increase or decrease. A calculator is not necessary.

1. The value was $100 and is now $113.
2. The population was 400 and is now 500.
3. The price was $50 and is now $54.
4. The number of deaths was 10 and is now 12.
5. The number of amoebas was 60 and is now 120.
6. The price was $40 and is now $36.
7. The value of a game is ⁻$1.75. How much should Dover expect to lose if he plays the game 40 times?
8. Roarke played a game 50 times and was $30 ahead. What was his average win?
9. Dino rolls a standard die once. If it lands on 6, he wins $10. Otherwise, he loses $2. Determine the value of the game.
10. Klarner rolls a standard die once. If it is 5 or 6, she wins $100. Otherwise, she loses $10. Determine the value of the game.
11. Of 456 students interviewed, 298 students thought Bryan Adams's name was Adam Bryans. What percent is this?

Copy and complete the Magic Squares.

12.

		3
0	2	4

13.

17	3	4	14
6		11	
5			

Determine the next two terms in the sequences.

14. 2, 1/2, 3, 1/3, 4, 1/4, 5…

15. a, 1, b, c, 2, d, e, f, 3, g, h, i…

GUINNESS RECORD®: The lowest height for a bar (flaming) under which a limbo dancer has passed is 6 1/8 inches off the floor by Mariene Raymond, 15, of Toronto, Canada on June 24, 1973. The record on roller skates (held by many people) is 7/8 inches lower than that.

16. What is the record on roller skates?

PROBABILITY
Lesson Plan 45

QUIZ

None

OBJECTIVES

Students will compare the theoretical value of a game with the experimental value of the game.

MATERIALS NEEDED

1. Experiment 8
2. Transparency 45
3. Three dice for the teacher

CLASS ACTIVITIES

Pass out Experiment 8 to each student and use the transparency to explain the game Chuck-A-Luck. This is a game that is played where gambling is legal. Each student will play the game 20 times. Make sure that the students have chosen their number before the three dice are rolled. In the outcome column, record the result of the roll of the dice. For example, 4, 6, 6. In the payoff column, the student should record his win or loss for that roll. After the game has been played 20 times, determine the average win/loss for the approximately 500 games played in your class. On the back of the experiment, have the students use a tree to determine the probabilities of 0, 1, 2, and 3 winners. The value of the game is:

3W	$1/216 \times \$3$
2W	$15/216 \times \$2$
1W	$75/216 \times \$1$
0W	$125/216 \times {}^-\$1$

Value Of The Game $\approx {}^-\$.08$

ASSIGNMENT

None

Chuck-A-Luck

The player chooses a number from 1 to 6. Three dice are rolled. If the player's number appears...

3 times	Win $3
2 times	Win $2
1 time	Win $1
0 times	Lose $1

EXPERIMENT 8

Chuck-A-Luck

Game	Your Bet	Outcome	Payoff	Cumulative
1.	_____	_____	_____	_____
2.	_____	_____	_____	_____
3.	_____	_____	_____	_____
4.	_____	_____	_____	_____
5.	_____	_____	_____	_____
6.	_____	_____	_____	_____
7.	_____	_____	_____	_____
8.	_____	_____	_____	_____
9.	_____	_____	_____	_____
10.	_____	_____	_____	_____
11.	_____	_____	_____	_____
12.	_____	_____	_____	_____
13.	_____	_____	_____	_____
14.	_____	_____	_____	_____
15.	_____	_____	_____	_____
16.	_____	_____	_____	_____
17.	_____	_____	_____	_____
18.	_____	_____	_____	_____
19.	_____	_____	_____	_____
20.	_____	_____	_____	_____

Results

	Me	Class
Number of bets	_____	_____
Total profit	_____	_____
Profit per bet	_____	_____

PROBABILITY
Lesson Plan 46

QUIZ   Quiz 18

OBJECTIVES   Students will be able to determine the value of a game when
                        Pascal's Triangle is used to determine the probabilities of
                        the outcomes.

MATERIALS NEEDED   1. Quiz 18
                        2. Transparencies 46A and 46B
                        3. Problem Set 46

CLASS ACTIVITIES   Transparency 46A gives information taken from the Fort Wayne
                        Journal-Gazette of Sunday, December 1 concerning a fund raiser
                        sponsored by the Big Brothers/Big Sisters of Whitley County. The
                        value of the game is ⁻$10.15. Transparency 46B gives an example of
                        the new material.

ASSIGNMENT   Problem Set 46

PROBABILITY QUIZ 18

Name _____

Complete the Chuck-A-Luck chart.

Game	Your Bet	Outcome	Payoff	Cumulative
1	3	1-1-4	_____	_____
2	4	4-4-6	_____	_____
3	1	2-3-5	_____	_____
4	1	3-4-6	_____	_____
5	5	3-3-4	_____	_____

2. The price of a video increased from $20 to $23. Determine the percent increase.

3. Only 7 of 25 interviewed students were happy with their hair color. What percent is that?

4. Of 60 students interviewed, only 10% had finished their Christmas shopping. How many had finished their shopping?

PROBABILITY 46A

Big Brothers/Big Sisters Raffle

2000 tickets at $50 each...

Prizes

1st	$74,500 home
2nd	$2,500 cash
3rd	$1,250 cash

What is the expected value of the game?

How much profit does Big Brothers/Big Sisters expect?

PROBABILITY 46B

Pascal and Expected Value

Four coins are tossed.

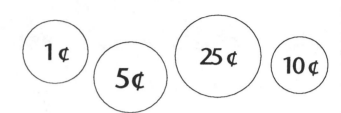

Game 1:

Four tails, win $50.

Otherwise, lose $3.

Should you play?

Game 2:

Four tails, win $100.

Three tails, win $50.

Otherwise, lose $20.

Should you play?

PROBABILITY
Problem Set 46

1. Malonza draws one card from a standard deck. If it is a ♦, he wins $50. Otherwise, he loses $10. Determine the value of the game.

2. Crowther draws one card from a standard deck. If it is an Ace, she wins $100. Otherwise, she loses $10. Determine the value of the game.

To work problems 3 and 4, begin by writing out the third row of Pascal's Triangle.

3. Three coins are tossed. If all land heads, Modesty wins $60. Otherwise, she loses $10. Determine the value of the game.

4. Three coins are tossed. If all land heads or all land tails, Lela wins $60. Otherwise, she loses $20. Determine the value of the game.

Calculate the following mentally.

5. 50% of 46   6. 10% of 70   7. 25% of 60   8. 5% of 40

9. 15% of 40

Calculate the following percent increase/decrease mentally.

Item	Original Cost	New Cost
10. Peanut	$100	$145
11. Grape	$50	$47
12. Grapefruit	$20	$40
13. Grape-Vine	$400	$396
14. Grape juice	$18	$27

The probability that a grapefruit is rotten is 12%. How many rotten grapefruit should Tabor expect to find if he buys

15. 100?   16. 50?   17. 800?   18. 548?

19. In a bunch of grapes, 16 were rotten and 47 were not. What percent were rotten?

20. Of 689 students interviewed, 4% ate Plum-Nuts for breakfast. How many ate Plum-Nuts?

21. Copy and complete the Magic Square.

8	1	$1\frac{1}{2}$	$6\frac{1}{2}$
	$5\frac{1}{2}$		4
	7		$\frac{1}{2}$

PROBABILITY
Lesson Plan 47

QUIZ   None

OBJECTIVES   Students will be able to determine the value of a game where the
                        game involves a tree with unequal length branches. Students will
                        be able to use a calculator to determine the percent increase
                        or decrease.

MATERIALS NEEDED   1. Transparencies 47A and 47B

                        2. Problem Set 47

CLASS ACTIVITIES   Transparency 47A sets up the situation. Have the students draw the
                        tree. Keep in mind that the game is over as soon as the WIN slip is
                        chosen. It is interesting that the probability of winning in one draw
                        (1/4) is the same as winning in two, three, or four draws. The value
                        of the game is $46.

                        Use transparency 47B to give examples of percent increase using
                        a calculator.

ASSIGNMENT   Problem Set 47

PROBABILITY 47A

Askew and The Lottery

A bowl contains three slips marked Loser and one marked Winner. Askew draws one at a time until he draws the Winner. The Payoff is as follows:

Winner on Draw One $100

Winner on Draw Two $60

Winner on Draw Three $20

Winner on Draw Four $4

How much can Askew expect to win?

PROBABILITY 47B

Percent Increase/Decrease (calculator)

The population increased from 17,876 to 19,432.

The price decreased from $45 to $38.

PROBABILITY
Problem Set 47

The probability that the drive-thru window at the local burger joint will fill Tedrow's order correctly is 7%. How many correct orders should Tedrow expect if he drives through

1. 100 times?   2. 500 times?   3. 74 times?   4. 692 times?

The value of a game is $4.76. How much should Yatrof expect to win if he plays the game

5. 2 times?   6. 484 times?

7. Dawson plays a game twice and wins a total of $5. What was his average win?

8. Ingrid plays a game 7 times and loses a total of $6. What was her average loss?

9. Write out the fifth row of Pascal's Triangle.

Five coins are tossed.

10. What is the probability that all coins land heads?

11. What is the probability that they don't all land heads?

12. If all five of the tossed coins land heads, Katandi wins $160. Otherwise she loses $5. Determine the value of the game.

13. If the number of heads showing equals the number of tails showing on the five tossed coins, Ignant wins $1,000,000,000. Otherwise he loses $1. What is the value of this game?

Calculate the percent increase/decrease. Use a calculator twice.

		Population	
	Town	1990	1991
14.	Nosuch City	0	0
15.	Pepper Lake City	138	147
16.	Old York	25	23
17.	Sinsinatti	43	50
18.	Deweyville	145	290

A box contains 87 widgets. Twenty of them are defective.

19. What percent are defective?   20. What percent are not defective?

21. How many are not defective?

22. Beket draws a widget at random. What is the probability that it is defective? Write the answer as a whole percent.

23. Becket draws two widgets without replacement. What is the probability that both are defective? Write the answer as a fraction.

Write the last term in each of the following sequences.

24. S, M, T, W, T, F, ____   25. J, F, M, A, M, J, J, A, S, O, N, ____

PROBABILITY
Lesson Plan 48

QUIZ

Quiz 19

OBJECTIVES

Students will determine the value of various bets that can be made in the game of Roulette. Then each student will play 20 games to compare the theoretical value of the game with the experimental value.

MATERIALS NEEDED

1. Quiz 19
2. Transparencies 48A and 48B
3. Experiment 9 (Transparency of this also)
4. Deck of cards numbered from 00, 0, 1–36 or some way of simulating the Roulette wheel.
5. Problem Set 48

CLASS ACTIVITIES

This lesson takes two days.

Transparency 48A describes a Roulette wheel. In Roulette, the wheel is spun and a marble is dropped and eventually comes to rest in one of the slots. Transparency 48B lists some of the bets can be made. Calculate the value of some of the bets. Note that 0 and 00 are neither odd or even, nor red or black. I believe that all bets have an expected value of about ⁻5¢. For example, red/black:

$$P(win) = 18/38 \times \$1$$
$$P(lose) = 20/38 \times {}^-\$1$$
$$Value = {}^-2/38 \approx {}^-5¢$$

Checking 0, 00:

$$P(win) = 2/38 \times \$17$$
$$P(lose) = 36/38 \times {}^-\$1$$
$$Value = {}^-2/38 \approx {}^-5¢$$

(continued on next page)

PROBABILITY
Lesson Plan 48

Next, give each student a copy of Experiment 10. In the Bet column, the student places one of the six bets listed at the top of the page. The Payoff column indicates the payoff as indicated at the top of the page. For example, a bet of Red has a payoff of 1:1. Students can only bet $1 each time. A payoff of 1:1 means the student wins $1 or loses $1. A payoff of 6:1 means a win of $6 or a loss of $1. Next, the teacher simulates a spin by drawing a card from the prepared deck. Numbers can be written in the appropriate colors on 3 x 5 cards. Students record the result: for example, *12,R* and then whether this was a win or loss. The profit is then recorded using the Payoff column. The Cumulative column keeps track of how they stand. Note that the most a student can lose is $20 and the most that they can win is $700 by betting the right number all 20 times. *Caution:* Make sure that students place their bets before you draw a card. After twenty spins, calculate the experimental value of the game for the students and the class.

ASSIGNMENT   Problem Set 48

PROBABILITY QUIZ 19

Name_____

1. There are five marbles, four green and one red, in a jar. Temklin draws one. If it is the red one, she wins $25. If it is a green one, she loses $7. What is the value of this game?

2. The value of a game is ⁻$.89. Tilker plays it 35 times. How much should he expect to lose?

3. Ramish played a game 12 times and won $16. What was her average win per game?

4. The number of unemployed rose from 50 to 57. What is the percent increase in unemployment?

PROBABILITY 48A

Roulette Wheel Description

38 slots: 0, 00, 1–36.

Green   0, 00

Red   1, 3, 5, 7, 9, 12, 14, 16, 18, 19, 21, 23, 25, 27, 30, 32, 34, 36

Black   2, 4, 6, 8, 10, 11, 13, 15, 17, 20, 22, 24, 26, 28, 29, 31, 33,

PROBABILITY 48B

Roulette

The following chart lists some of the bets that can be made in roulette and the payoff.

Bet	Payoff
red/black	1 to 1
odd/even	1 to 1
1–12/13–24/25–36	2 to 1
0, 00, 1, 2, 3	6 to 1
1–3/4–6/7–9/…	11 to 1
0,00	17 to 1
single number	35 to 1

Which of these is the best bet?

EXPERIMENT 9

Roulette

Bets/Payoff

Odd/Even; Red/Black:	1 to 1
1–18; 19–36:	1 to 1
1–12; 13–24; 25–36:	2 to 1
0, 00, 1, 2, 3:	6 to 1
Any particular number:	35 to 1

Spin	Bet	Payoff	Result	W/L	Profit	Cumulative
1.	_____	_____	_____	_____	_____	_____
2.	_____	_____	_____	_____	_____	_____
3.	_____	_____	_____	_____	_____	_____
4.	_____	_____	_____	_____	_____	_____
5.	_____	_____	_____	_____	_____	_____
6.	_____	_____	_____	_____	_____	_____
7.	_____	_____	_____	_____	_____	_____
8.	_____	_____	_____	_____	_____	_____
9.	_____	_____	_____	_____	_____	_____
10.	_____	_____	_____	_____	_____	_____
11.	_____	_____	_____	_____	_____	_____
12.	_____	_____	_____	_____	_____	_____
13.	_____	_____	_____	_____	_____	_____
14.	_____	_____	_____	_____	_____	_____
15.	_____	_____	_____	_____	_____	_____
16.	_____	_____	_____	_____	_____	_____
17.	_____	_____	_____	_____	_____	_____
18.	_____	_____	_____	_____	_____	_____
19.	_____	_____	_____	_____	_____	_____
20.	_____	_____	_____	_____	_____	_____

	Me	Class
Number of bets	_____	_____
Total profit	_____	_____
Profit per bet	_____	_____

PROBABILITY
Problem Set 48

1. Write out the first five rows of Pascal's Triangle.

2. Rumpkin spins the spinner five times. The payoff for the game is as follows:

   Five wins, win $200.
   Otherwise, lose $6.
   Determine the value of the game.

3. Tryshn spins the spinner five times. The payoff for the game is as follows:

   Five wins, win $100.
   Four wins, win $50.
   Otherwise, lose $15.
   Determine the value of the game.

4. A bag contains three red and seven blue marbles. One is chosen and if it is red, Ystrin wins $50. Otherwise, she loses $10. Determine the value of the game.

5. A gizmo that previously cost $50 now costs $65. What was the percent increase in price?

Dr. Wate Loss told overweight Orville to lose 10% of his weight. Orville weighed 430 pounds.

6. How much should he lose?

7. What should his new weight be?

Of 674 teenagers interviewed, only 632 believed in Santa Claus.

8. What percent believed in Santa?

9. What percent did not believe in Santa?

10. How many did not believe in Santa?

Copy and complete the Magic Squares.

11.

9	2	$2\frac{1}{2}$	$7\frac{1}{2}$
	$6\frac{1}{2}$	6	
$5\frac{1}{2}$			
	8		

12.

3		1
-2		
-1		

GUINNESS RECORD®: The record for the fastest speed for a land snail was recorded by a a snail named Verne at West Middle School in Plymouth, MI on February 21, 1990 when he traveled slightly over 12 inches in 2 minutes and 13 seconds.

13. At that rate, how long would it take Verne to crawl 100 yards?

14. At that rate, how long would it take Verne to travel a mile? One mile = 5,280 feet. Give your answer in terms of days.

PROBABILITY
Lesson Plan 49

QUIZ   Quiz 20

OBJECTIVES   Given a set of numbers, students will be able to determine the
                        average (mean) of the set of numbers.

MATERIALS NEEDED   1. Quiz 20
                        2. Transparency 49
                        3. Problem Set 49

CLASS ACTIVITIES   The concept of computing the average of a set of numbers is
                        introduced by means of wins and losses in a game.

ASSIGNMENT   Problem Set 49

PROBABILITY QUIZ 20

Name _____

1. Complete the chart for three games of Roulette.

Spin	Bet	Payoff	Result	W/L	Profit	Cumulative
1	16–18	11:1	18,R	_____	_____	_____
2	odd	1:1	8,B	_____	_____	_____
3	27	35:1	20,B	_____	_____	_____

Number of games _____

Total profit/loss _____

Average profit/loss _____

2. Thirty-eight Eskimos used Chap-Stick and 12 did not. What percent used Chap-Stick?

3. Of 300 beach bums interviewed, 25% used sun-block. How many used sun-block?

4. Last year 400 mountain men used razors. This year 420 used them. What is the percent increase?

PROBABILITY 49

Gaby and the Game

Gaby played a game five times and won a total of $15. What was her average win per game?

Eunice played a game five times and won $4, $1, $1, $5, and $4. What was her average win per game?

Kimball played a game four times. Her profits were $5, ⁻$3, ⁻$4, and $6. What was her average win per game?

Terkel played a game five times and her average win was $3.00. Her profit on the first four games was $2, $2, $4, and $1.50. How much did she win on the last game?

PROBABILITY
Problem Set 49

1. Vanslow had quiz scores of 8, 7, 3, and 8. Determine her average score to the nearest tenth.

2. On the next quiz Vanslow, got caught cheating and received a score of 0. What is her new average?

3. What score must Vanslow get on her next quiz to get her average up to 6?

4. Merton had an average score of 65 on two tests. One of his test scores was 80. What was his score on the other test?

5. A bowl contains seven red, two white, and one blue marble. Greer draws one. If it's blue she wins $100, if it's white she wins $50, otherwise, she loses $30. Determine the value of the game.

A bowl contains three red and two white marbles. Yoda chooses two without replacement.

6. P(red,red) = ?

7. If Yoda chooses two red marbles, he wins $50. Otherwise, he loses $10. Determine the value of the game.

Sheba scored 56 on the first test and 70 on the second test.

8. What was the percent increase in her scores?

9. What was her average score?

10. What might you call a mouse that weighed 505 pounds (Enter 505), gained 16 pounds every year (Enter x 16) and then lost both ears (Enter − 2)?

Thadeus bought a coat that originally cost $480. However, it was on sale for 10% off.

11. How much money did Thadeus save?

12. How much did the coat cost on sale?

13. How much sales tax (5%) did Thadeus have to pay?

14. Thadeus paid with four $100 bills, a $50 bill and a $10 bill. How much change did he receive?

On Monday, 34 students were absent and 698 were present at school.

15. How many students are enrolled at the school?

16. What percent were present?   17. What percent were absent?

GUINNESS RECORD®: The heaviest twins were Billy and Benny McCrary of Henderson, North Carolina. In November 1978, Billy weighed 743 pounds and Benny weighed 723 pounds.

18. What was their average weight?

19. What percent of their total weight did Billy weigh?

PROBABILITY
Lesson Plan 50

QUIZ   None

OBJECTIVES   No new material is presented in this lesson.

MATERIALS NEEDED   1. Transparency 50
                        2. Problem Set 50

CLASS ACTIVITIES   Transparency 50 gives information concerning the Daily Three and Daily Four Lottery games played in Indiana. When played straight the player must have the numbers in the correct order. The probability of winning is 1/1000 with a payoff of $500. Hence the value of the game (Daily Three played straight) is ⁻49.9¢.

When the bet is boxed, any of the six combinations of the chosen numbers is a winner. That is, if you choose 567, other winners are 576, 675, 657, 765, and 756. However, the value of this game is ⁻51.4¢.

In Daily Four, the value of the straight game is ⁻49.99¢ and the value of the boxed game (24 winning combinations) is ⁻51.76¢. Note that the player must declare whether he is playing the boxed or straight game. From these calculations, one can conclude that the best game is The Daily Three played straight.

ASSIGNMENT   Problem Set 50

PROBABILITY 50

Daily Three / Daily Four

The Daily Three

Choose any three digit number from 000–999.

The Payoff

Bet it straight $500

Bet it boxed $80

The Daily Four

Choose any four digit number from 0000–9999.

The Payoff

Bet it straight $5000

Bet it boxed $200

PROBABILITY
Problem Set 50

GUINNESS RECORD®: The record for killing rats by a dog is held by a bull terrier named Jenny Lind. On July 12, 1853 she killed 500 rats in 1 hour 36 minutes in Liverpool, England.

1. How many minutes did it take her to kill the rats?
2. What was the average number of rats she killed per minute?

Fabion tossed four pennies and was paid as follows:

    All tails, win $50
    Otherwise, lose $4

3. Write out the fourth row of Pascal's Triangle.
4. What is the probability that Fabion will win? P(4T) = ?
5. What is the probability that Fabion will lose?
6. Determine the value of the game.
7. How much should he expect to lose if he played the game 40 times?
8. Novell tosses a standard die. If it lands showing "1," he is paid $5. Otherwise he loses $1. Determine the value of the game.

Of 46 thespians interviewed, about 35% preferred Maxgood House Coffee.

9. How many preferred Maxgood House?   10. How many didn't prefer Maxgood House?
11. What percent didn't prefer Maxgood House?

Of 47 architects interviewed, 40 used Bac Clac pens.

12. What percent used Bac Clacs?
13. Three more architects were interviewed and all of them used Bac Clacs. What percent of the 50 use Bac Clacs?
14. The average weight of a player on the Carlton Curling Team is 46 pounds. If there are 15 players on the team, what is there total weight?
15. The average IQ of five English teachers was 90. Four of the teachers had IQ's of 110. What was the IQ of the fifth teacher?

On four tests Ima Student scored 89, 82, 70, and 93. Unfortunately, the teacher recorded the 93 as a 39.

16. According to the teacher, what was Ima's average?
17. What was Ima's actual average?
18. Last year Cream Abdull Gibar made $4,500. This year he made $5,000. Determine the percent increase.
19. Complete the Magic Square to the right.

9	2	$2\frac{1}{2}$	$7\frac{1}{2}$
	$6\frac{1}{2}$	6	
$5\frac{1}{2}$			
	8		

PROBABILITY
Test 5

Do not write on this test. Do all work on scratch paper and put the answer on the appropriate space on the answer sheet.

GUINNESS RECORD®*:* The first woman to lift more than her body weight with one arm was Cammie Lynn Lusko who lifted 101.9% of her body weight on May 21, 1983 in Milwaukee, Wisconsin.

1. She weighed 128.5 pounds. How much did she lift?

The value of a game is $1.75. How much should Travis expect to win if he plays the game

2. 2 times?   3. 45 times?

Pentour plays a game five times. Determine the average win if she wins a total of

4. $15.00.   5. $4.00.

Tupwig throws a dart at the dartboard.

6. P(A)   7. P(not A)

8. If the dart lands in A, Tupwig wins $3.00. Otherwise, he loses $1.00. Determine the value of the game.

A deck of nine cards has the letters C, H, R, I, S, T, M, A, S written on them. Yoplait chooses one at random.

9. P(C)   10. P(not C)

11. If the card is a C, Yoplait wins $10.00. Otherwise, she loses $1.00. Determine the value of the game.

Ewint tosses four coins.

12. Write out the fourth row of Pascal's Triangle.   13. P(2H,2T)

14. Ewint wins $50.00 if the coins all land heads or all land tails. Otherwise, he loses $10.00. Determine the value of the game.

15. The probability of winning $10,000 in Litl Lotto is 1/5000. What is the probability of not winning?

Of 600 attorneys surveyed, only 10% admitted having seen Rudolph the Red-Nosed Reindeer.

16. How many had seen Rudolph?   17. How many had not seen Rudolph?

18. What percent had not seen Rudolph?

PROBABILITY
Test 5

In a survey of kindergartners, 43 could name Santa's eight reindeer and seven could not.

19. How many students were surveyed?   20. What percent could name eight reindeer?

21. In a survey of 598 truck drivers, 79% knew all of the words to "Silent Night." How many knew all of the words.

Of 975 electricians, only 163 had "seen mommy kissin' Santa Claus."

22. What percent had seen this event?

23. What percent had not seen mommy kissing Santa?

24. How many had not seen mommy kissing Santa?

25. The amusement park has a new attraction. You begin by climbing 25 feet (Enter 25). Then you enter any of seventeen pathways (Enter x 17) and travel 71 feet to the other end (Enter x 71). What is the attraction?

26. The population of Lakattle increased from 50 to 56. What was the percent increase?

27. Last year 415 people attended the Bouncing Rocks concert. This year 830 people attended. What was the percent increase?

28. A pineapple that cost $60.00 last year only costs $45.00 this year. What was the percent decrease?

29. Last week 765 people lost their temper. This week an additional 35 people lost their temper. What is the percent increase?

The story of the dwarfs:

30. The weights of the five dwarfs are 6, 7, 11, 9, and 7 pounds. Find the average weight.

31. Dumbo the Dwarf joins the first five dwarfs. Dumbo weighs 20 pounds. Find the average weight of the six dwarfs.

32. Doxille the Dwarf joins the six. The average weight of the seven dwarfs is now 11 pounds. How much does Doxille weigh?

33. The average height of the seven dwarfs is 22 inches. What is their total height?

The probability of a white Christmas in Georgia is 4%. How many white Christmases should you expect to see in Georgia in

34. 100 years?   35. 50 years?   36. 65 years?

Write the next two terms in each sequence.

37. 1, ⁻3, 5, ⁻7, 9, ⁻11…   38. m, E, r, R, y, C, h, R, i, S, t, M…

Complete the Magic Squares on the answer sheet for problems 39 and 40.

PROBABILITY

Test 5 Answer Sheet

Name _____

1. _____

2. _____

3. _____

4. _____

5. _____

6. _____

7. _____

8. _____

9. _____

10. _____

11. _____

12. _____

13. _____

14. _____

15. _____

16. _____

17. _____

18. _____

19. _____

20. _____

21. _____

22. _____

23. _____

24. _____

25. _____

26. _____

27. _____

28. _____

29. _____

30. _____

31. _____

32. _____

33. _____

34. _____

35. _____

36. _____

37. _____

38. _____

39.

4	$\frac{1}{2}$	
	$2\frac{1}{2}$	
		1

40.

13	-1	0	10
	8		
		3	9

PROBABILITY
Lesson Plan 51

QUIZ   None

OBJECTIVES   Given a set of data, students will be able to determine its median. Also, students will realize that in some cases, the median is a better representation of the data than the mean.

MATERIALS NEEDED   1. Transparency 51
                             2. Problem Set 51

CLASS ACTIVITIES   One way of starting the class is to determine the average age of everyone in the classroom. Since one person is much older than the rest, the average is misleading. This shows the need for another average, namely the median. Transparency 51 defines the median and gives some examples. Also, calculate the median age of the people in the classroom. Now that we have two "averages," we will use the word *mean* to describe our first average. Also if the word average is used, it will mean *mean*.

ASSIGNMENT   Problem Set 51

PROBABILITY 51

The Median

Arrange the numbers from large to small (or small to large). The middle number is the *median*. If there is no middle number, the median is the average of the two middle numbers.

Example 1: 5, 9, 6, 6, 3, 9, 5, 2, 6

Example 2: 3, 8, 8, 6, 9, 7

PROBABILITY
Problem Set 51

The following indicate the amount of time that Baden spent waiting in line to buy popcorn on fifteen trips to the concession stand: 14, 1, 9, 7, 5, 16, 1, 9, 2, 8, 4, 8, 0, 9, 12.

1. Find the mean number of minutes spent waiting.

2. Find the median number of minutes spent waiting.

The following indicate the number of used cars sold by ten salespersons at Anabel's Auto Arena: 63, 75, 90, 69, 51, 71, 94, 49, 63, 84

3. Find the mean number of cars sold.   4. Find the median number of cars sold.

The following indicates the number of automobiles owned by seven families: 4, 6, 4, 7, 9, 8, 7

5. Find the mean number of cars owned.   6. Find the median number of cars owned.

When Shalena Student was working problems 5 and 6, he accidently put down 99 instead of 9.

7. Calculate the mean using 99 instead of 9.

8. Calculate the median using 99 instead of 9.

9. Brutus scored 70 on each of his first three tests. He scored 38 on his fourth test. Determine his mean score.

10. Julius scored 70 on each of his first three tests. What must he score on his next test to raise his average to 75?

On the Carlton Curling team, the three girls had an average weight of 80 pounds and the 7 boys had an average weight of 98 pounds.

11. What was the total weight of the girls?   12. What was the total weight of the boys?

13. What was the total team weight?   14. How many players were on the team?

15. What was the average weight of a member of the team?

GUINNESS RECORD®: On July 16, 1989, George Uhrin, of Houston, TX memorized a random sequence of thirty standard decks of cards that had been shuffled together. He looked at the cards only once and made only two errors.

16. How many cards were in the 30 decks?

17. What percent of his answers were correct? (nearest tenth)

18. Last week Phil T. Rich made $800.00. This week he made $950.00. Determine his percent increase in salary.

19. Marmaduke ate 46 of the 67 bones he had buried. What percent of the bones did he eat?

20. Only 17% of the children in Spoildville liked the gifts that Santa left them. There are 458 children in Spoildville. How many liked their gift?

PROBABILITY
Lesson Plan 52

QUIZ   None

OBJECTIVES   Given a set of data, students will be able to exhibit it in a frequency distribution. A tally chart is needed before the bar chart (frequency distribution) can be made. Also, students will be able to determine the mean and median from the frequency distribution.

MATERIALS NEEDED   1. Transparency 52
                     2. Problem Set 52

CLASS ACTIVITIES   One way to introduce the topic is to gather data from your class and make a frequency distribution. Suggested data:

                     Number of TV sets
                     Number of brothers/sisters
                     Number of telephones

                     Transparency 52 gives a second example. Students should be able to determine the mean and median from the frequency distribution.

ASSIGNMENT   Problem Set 52

PROBABILITY 52

Corporal Cop and Spitting

The following gives the number of tickets issued by Corporal Cop for spitting on the sidewalk over the past 20 days:

6	3	2	4	6
3	3	5	2	6
4	4	6	4	4
3	4	6	2	3

Make a tally chart and a frequency distribution. Then determine the median and mean.

PROBABILITY
Problem Set 52

BROTHERS/SISTERS: Thirty students were asked how many brothers and sisters they had. The results were as follows:

```
0 3 3 1 6 5 0 1 1 2
2 2 4 0 1 4 4 3 1 5
0 1 4 2 2 3 2 3 4 1
```

1. Make a tally chart and then draw a bar chart (frequency distribution) of this information.
2. What was the mean number of brothers and sisters?
3. What was the median number of brothers and sisters?
4. What percent of the students have three brothers and sisters?
5. What percent of the students have more than the mean number of brothers and sisters?

STUDENT ABSENCES: The bar chart shows how many days a group of students were absent last week.

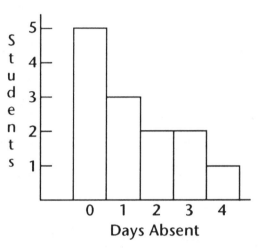

6. How many students were absent 3 days?
7. How many students had perfect attendance?
8. How many students were included in the survey?
9. What percent of the students were absent 4 days?
10. Determine the mean number of absences.
11. Determine the median number of absences.

MINIATURE GOLF: Benjamin and Franklin played nine holes of miniature golf. The results are as follows:

Hole	1	2	3	4	5	6	7	8	9
Benjamin	4	3	5	2	2	4	3	5	2
Franklin	4	2	5	1	2	3	3	4	8

12. Determine the mean score for Benjamin.
13. Determine the mean score for Franklin.
14. Determine the median score for Benjamin.
15. Determine the median score for Franklin.
16. On the last test, 12 right-handed students averaged 70 and three left-handed students averaged 90. What was the average for the 15 students?

GUINNESS RECORD®: The greatest number of times a pancake has been tossed in two minutes is 281, by Judith Aldridge on February 27, 1990.

17. At that rate, how many times would she toss it in an hour?

PROBABILITY
Lesson Plan 53

QUIZ

Quiz 21

OBJECTIVES

Given a set of data or a frequency distribution, students will be able to determine the range of the data.

MATERIALS NEEDED

1. Quiz 21
2. Transparency 53
3. Problem Set 53

CLASS ACTIVITIES

Transparency 53 introduces the concept. The mean and median of the pulse rates of Egore and Elvira are about the same. However, Elvira's rate is much more erratic. The range (the difference between the high and low) tells how the data is spread out. The Quiz Analysis gives more practice in determining mean, median, and range from a frequency distribution.

ASSIGNMENT

Problem Set 53

PROBABILITY QUIZ 21

Name _____

Willard the Whiz scored the following on five biology quizzes: 10, 5, 5, 9, 6.

1. Determine the mean (average score).

2. Determine the median score.

Brenda the Brain scored the following on eight biology quizzes: 10, 6, 8, 8, 3, 9, 9, 4.

3. Determine the mean (average score).

4. Determine the median score.

5. Blodyn the Bum scored 5 and 6 on his first two quizzes. What must he score on the next quiz to get his average to 7?

PROBABILITY 53

Who's Healthier?

The following are pulse rates for Egore and Elvira taken at three times.

Egore	72 76 74
Elvira	59 92 71

	Egore	Elvira
mean		
median		
range		

Quiz Analysis

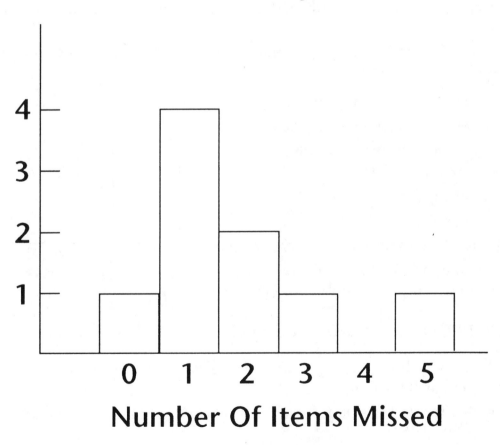

Number Of Items Missed

PROBABILITY
Problem Set 53

GUINNESS RECORD®: On June 30, 1989, John Kenmuir licked and affixed 328 stamps in four minutes at George Square Post Office, Glascow, Scotland.

1. How many did he lick and affix per minute?
2. How many did he lick and affix per second? (to the nearest tenth)
3. If each stamp cost 29¢, how much did the stamps cost?
4. If the stamps were in a roll, and each stamp was one inch long, how many feet long would the unrolled roll have been? (to the nearest tenth)

Nine students received the following numbers of pieces of mail: 1, 3, 2, 0, 1, 5, 2, 1, 3

5. Determine the mean number of pieces of mail received.
6. Determine the median number of pieces of mail received.
7. Determine the range of the number of pieces of mail received.

The average monthly rainfall in Drenchville was 14 inches.

8. How many inches of rain does Drenchville receive in one year?
9. What is the least amount of rain Drenchville could have received in one month?
10. What is the largest amount of rain Drenchville could have received in one month?
11. Keziah averaged 89 on two tests. If her first test score was 86, what was her second?

THE TERRIFIC TYPISTS: The frequency distribution shows the number of errors made by students in a typing class.

12. How many students made three errors?
13. What was the greatest number of errors?
14. How many students were in the class?
15. Find the mean number of errors.
16. Find the median number of errors.
17. Find the range of the error.
18. What percent of the students made three errors?
19. The oldest swimmer was 97 and the range of the ages was 67. How old was the youngest swimmer?

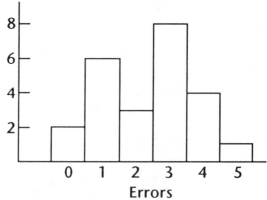

AN UNINTERESTING CLASS: In a certain class, all 23 students have an IQ of 110.

20. What is the mean IQ?   21. What is the median IQ?
22. What is the range of the IQ's?
23. The year 1991 is a PALINDROME. It reads the same forward as backward. What year is the next palindrome?

PROBABILITY
Lesson Plan 54

QUIZ

None

OBJECTIVES

Students will determine the mean, median, and range of the number of draws it takes before an Ace is drawn from a standard deck.

MATERIALS NEEDED

1. Transparency 54
2. Experiment 10 (transparency)
3. A table of random numbers for each student
4. A deck of cards for each pair of students (optional)

CLASS ACTIVITIES

Transparency 54 is used in discussing the last problem in Problem Set 53. It takes about 30 minutes to run the experiment. I suggest that you run two or three trials with the students in order to make sure that they understand the process.

ASSIGNMENT

None

PROBABILITY 54

Palindromes

radar   mom   dad   level

Dennis sinned.

Madam, I'm Adam.

A man, a plan, a canal. ___ ___ ___ ___ ___ ___.

Was it a car or ___ ___ ___ ___ ___ ___ ___ ___?

GUINNESS RECORD®: The longest palindromic composition is one of 100,000 words by two Englishmen. It begins "Al, sign it, Lover!…" and ends "…revolting, Isla."

EXPERIMENT 10

How Long To Get An Ace?

In this experiment we'll determine how many cards must be dealt from a standard deck of 52 cards before an Ace is drawn. The fewest number of cards that could be drawn is _____ and the greatest number is _____.

Use a table of random numbers to simulate the experiment. Cards are represented by numbers 1–52. Numbers 1, 14, 27, and 40 represent Aces. 00 and numbers larger than 52 are disregarded. Once a number is chosen on a trial it is disregarded any other time it is chosen during the trial.

Run the experiment 15 times and complete the bottom of the sheet.

Trial	Outcomes	Number
1		
2		
3		
4		
5		
6		
7		
8		
9		
10		
11		
12		
13		
14		
15		

The mean number of cards drawn in order to get an Ace was _____.

The median number of cards drawn in order to get an Ace was _____.

The range in the number of cards drawn in order to get an Ace was _____.

PROBABILITY
Random Number Table

19 61 39 64 14	27 84 17 75 62	03 90 68 74 30	00 04 15 55 19	18 57 78 00 70
11 28 66 19 00	08 56 74 06 47	70 42 71 44 09	48 12 28 38 36	29 39 40 98 50
77 73 60 36 25	03 29 46 87 56	45 71 95 17 69	64 94 06 83 93	81 84 78 47 81
35 86 31 86 81	34 03 06 27 03	26 65 62 44 45	36 90 32 34 65	23 53 84 33 11
21 30 01 66 66	45 27 59 33 67	07 11 56 78 70	89 16 25 55 07	33 37 56 71 18
03 98 88 37 59	46 62 98 06 91	83 64 67 68 09	82 59 79 60 59	37 74 83 24 19
06 53 83 05 47	63 64 33 30 36	56 75 52 16 39	54 94 04 41 83	90 20 77 22 72
14 53 66 35 63	43 98 51 37 63	46 08 43 58 71	56 80 74 63 81	94 87 56 95 73
71 03 43 00 74	58 49 27 81 49	09 28 82 82 35	18 95 69 22 67	90 58 97 68 59
19 81 19 49 36	80 57 04 32 38	44 83 54 60 07	88 39 10 90 17	48 51 42 38 91
24 35 89 58 72	14 27 14 02 85	30 65 15 41 82	07 67 34 56 10	56 10 74 70 29
70 95 77 43 59	38 92 64 62 54	39 00 90 17 73	74 28 77 52 51	65 34 46 74 15
21 81 85 93 13	93 27 88 17 57	05 68 67 31 56	07 08 28 50 46	31 85 33 77 74
70 34 20 79 26	72 06 70 40 26	75 17 97 88 23	33 61 53 03 18	82 68 88 44 21
91 46 62 36 83	90 92 37 36 12	75 52 93 34 53	84 43 75 48 62	13 22 05 74 36
21 72 44 58 51	71 83 47 37 27	93 64 94 21 68	74 05 50 00 27	14 99 51 66 48
24 52 83 19 22	76 09 16 59 32	41 23 05 08 94	89 85 74 48 11	03 77 86 95 36
14 62 19 97 30	13 26 78 89 62	68 84 01 74 96	75 30 36 81 32	53 60 90 06 24
70 29 88 36 74	52 21 30 43 80	02 31 87 70 94	53 67 77 65 62	00 89 11 01 43
03 33 82 91 14	88 99 87 16 74	43 10 43 43 96	69 98 36 20 97	07 36 18 67 86
05 13 18 00 65	47 50 56 16 57	63 25 62 95 57	62 81 12 46 20	76 16 45 95 47
47 07 96 00 89	58 81 22 29 58	46 07 22 52 58	05 40 40 26 85	03 89 69 90 79
81 05 47 31 35	51 25 60 07 15	34 39 84 22 32	38 52 99 95 83	19 59 70 90 55
37 14 90 87 52	34 90 88 27 54	71 64 58 85 88	30 18 40 66 75	60 34 91 17 33
24 60 09 07 21	85 62 15 53 28	35 04 94 62 92	51 69 12 76 98	68 78 96 07 36
27 89 70 31 42	52 03 35 60 36	82 05 17 29 06	98 22 45 48 75	52 16 21 23 60
59 93 94 48 05	64 89 47 42 96	24 80 52 40 37	20 63 61 04 02	00 82 29 16 65
08 42 26 89 53	19 64 50 93 03	23 20 90 25 60	15 95 33 47 64	35 08 03 36 06
99 01 90 25 29	09 37 67 07 15	38 31 13 11 65	88 67 67 43 97	04 43 62 76 59
12 80 79 99 70	80 15 73 61 47	64 03 23 66 53	98 95 11 68 77	12 17 17 68 33
66 06 57 47 17	34 07 27 68 50	36 69 73 61 70	65 81 33 98 85	11 19 92 91
31 06 01 08 05	45 57 18 24 06	35 30 34 26 14	86 79 90 74 39	23 40 30 97
85 26 97 76 02	02 05 16 56 92	68 66 57 48 18	73 05 38 52 47	87 18 19 91
16 78 74 80 93	83 40 59 75 27	66 65 52 22 52	59 60 23 29 49	07 82 72 09
32 83 36 86 75	48 59 24 05 07	00 45 28 60 37	75 72 76 01 55	82 74 16 18

PROBABILITY
Lesson Plan 55

QUIZ

None

OBJECTIVES

Review mean, median, and range.

MATERIALS NEEDED

1. Transparency 55
2. Problem Set 55

CLASS ACTIVITIES

The following procedure always eventually leads to a palindome. Begin with any number. Reverse its digits and add to the original number. If the result is not a palindrome, repeat the process. For all *two digit* numbers, if the sum of their digits is less than 10, a palindrome will result in one addition. Other situations are as follows:

Sum Of Digits	Additions To Palindrome
10	2
11	1
12	2
13	2
14	3
15	4
16	6
17	24
18	6

The problems on the transparency verify these results.

ASSIGNMENT

Problem Set 55

PROBABILITY 55

Creating Palindromes

1. Choose any number.

2. Reverse its digits.

3. Add the new number to the original.
   If the result is not a palindrome, repeat
   the procedure. Eventually a palindrome
   will appear.

Examples

34   67   68   79

PROBABILITY
Problem Set 55

Four judges rated five breakfast cereals on a scale from 1–10 as follows:

Cereals	Scalia	O'Conner	Thomas	Rehnquist
White Ties	5	6	2	9
Gloomyos	8	8	8	9
Meely Oats	4	9	9	8
Flax Flakes	4	6	10	9
Soggies	4	5	8	7

Determine the *mean* score (nearest tenth) for:

1. White Ties   2. Gloomyos   3. Meely Oats   4. Flax Flakes   5. Soggies

Determine the median score for:

6. White Ties   7. Gloomyos   8. Meely Oats   9. Flax Flakes   10. Soggies
11. What was the "best" cereal if the mean score is used?
12. What was the "best" cereal if the median score is used?

Determine the median score given by:

13. Scalia   14. O'Conner   15. Thomas   16. Rehnquist
17. Who was the toughest judge according to the median score?

Determine the range of the scores given by:

18. Scalia   19. O'Conner   20. Thomas   21. Rehnquist
22. What judge was the most consistent according to the range?
23. To what group do these judges belong?

Use the "reverse and add" procedure to obtain a palindrome.

24. 48   25. 7

GUINNESS RECORD®: The largest watermelon grown was raised by B. Rogerson of Robersonville, North Carolina in 1988. It weighed 279 pounds.

26. If each person can eat 3 pounds of watermelon, how many people would it take to eat it?
27. The typical watermelon weighs about 25 pounds. How many typical watermelons is Rogerson's equivalent to?
28. About 93% of a watermelon is water. How many pounds of water was in Rogerson's watermelon?

PROBABILITY
Lesson Plan 56

QUIZ

Quiz 22

OBJECTIVES

Given a set of data, students will be able to organize it into an unordered stem and leaf plot and then an ordered stem and leaf plot. Then students will be able to determine the upper and lower quartiles.

MATERIALS NEEDED

1. Quiz 22
2. Transparency 56
3. Problem Set 56

CLASS ACTIVITIES

The unordered stem and leaf plot is made by using the tens digit for the stem and the ones digit for the leaves. Ordering the leaves creates an ordered stem and leaf plot. The results for the transparency are shown.

8	2
7	13158
6	9210014
5	73346863278
4	9

8	2
7	11358
6	0011249
5	23334667788
4	9

The median is 60. To calculate the upper quartile, a number above which $\frac{1}{4}$ of the data lies, simply find the median of the upper half of the data. To calculate the lower quartile, do the same with the lower half of the data.

The upper quartile is 70 and the lower quartile is 55.

ASSIGNMENT

Problem Set 56

PROBABILITY QUIZ 22

Name _____

The frequency distribution shows the results of a survey of the number of classes students take.

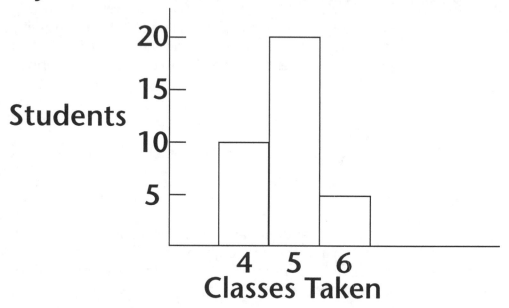

1. How many students took 4 classes? _____

2. How many students were surveyed? _____

3. What percent of the students took 4 classes? _____

4. What was the greatest number of classes any student took? _____

5. What was the range of the classes taken? _____

6. What was the median number of classes taken?

   _____

7. What was the mean number of classes taken?

   _____

PROBABILITY 56

Paula The Police Officer

Using a radar gun, Paula determines the speeds of 25 cars.

57	53	53	71	73
54	69	56	58	49
56	53	52	82	62
61	60	71	75	60
57	61	58	78	64

1. Draw a stem-leaf plot.

2. Determine the range of the speeds.

3. Determine the median speed.

4. Determine the upper quartile.

5. Determine the lower quartile.

PROBABILITY
Problem Set 56

GUINNESS RECORD®: The largest collection of valid credit cards is owned by Walter Cavanaugh of Santa Clara, California. He owns 1,212 credit cards.

1. If each card allows him to charge $2,000 worth of merchandise, how much could he charge?
2. A standard credit card is 2.8 inches long. If the cards were lined up end to end, how long (in inches) would the line be?
3. How long would the line be in feet?
4. A stack of 32 cards is 1 inch tall. How many inches tall would Mr. Cavanaugh's stack be?
5. How many feet tall would his stack be?

The odometer on Odie's automobile reads 3437.

6. At what mileage will the odometer next be a palindrome?
7. How many miles does Odie need to travel before he reaches the next palindrome?
8. Yasir wants to set his digital alarm clock for a palindrome so that he gets up as close to 6:00 as possible. What time should he set his alarm for?

```
42 51 56 55
51 54 51 60
62 43 55 56
61 52 69 64
```

The table to the right gives the ages at inauguration of the 16 United States Presidents inaugurated in the 20th century.

9. Draw an unordered and then an ordered stem and leaf plot.
10. What was the range of their ages?   11. What was the median age?
12. What was the upper quartile age?

```
9 | 2 3 7 8
8 | 4 9
7 | 8 9
6 | 0 3
```

The stem and leaf plot to the right gives the test scores of Ms. Mag's first period drafting class.

13. How many students are in the class?   14. What was the high score on the test?
15. What was the low score?   16. What was the mean score?
17. What was the median score?   18. What was the upper quartile score?

BUBBLE GUM RATING: A group of students were asked to rate (on a scale of 1 to 10) a new brand of gum. The results are shown in the frequency distribution.

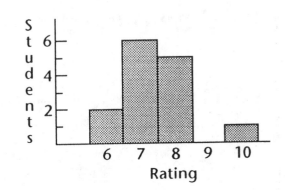

19. How many gave the gum a 7?
20. What was the lowest score given?
21. How many students took part in the survey?
22. What percent of students gave a score of 7?
23. What is the range of scores?
24. What is the median score?   25. What is the mean score?

PROBABILITY
Lesson Plan 57

QUIZ

None

OBJECTIVES

Given two sets of data, students will be able to compare them by drawing back-to-back stem and leaf plots.

MATERIALS NEEDED

1. Transparency 57a and 57b
2. Problem Set 57

CLASS ACTIVITIES

Transparency 57a gives information concerning the number of times students in two of a teacher"s classes could toot a trumpet. To compare the classes, the teacher places the scores on a back-to-back stem and leaf plot. (See below.)

1st Period		5th Period
	10	0 0 0
8 5 3 0	9	
6 5 5 5 4 2 0	8	1 2 5
9 8 8 8 5 0	7	0 2 5 5
8	6	0 2 5 6 9
	5	3 5 8
	4	0
9 2	3	

From this chart it is easy to see that the first period class did much better than the fifth period. One should also calculate the range, quartiles, and mean. Transparency 57b can be used to compare Larry Bird and Earvin Johnson in two categories. In both cases, Bird comes out on top. Does this make him a better player?

ASSIGNMENT

Problem Set 57

PROBABILITY 57A

First Or Fifth

Thadeus Twolips, the trumpet teacher, wanted to compare his first period with his fifth period class. He counted the number of times each student could toot the trumpet and recorded the results.

First Period

98 70 75 85 86 79 80 82 39 68 95 93 85
85 90 78 78 32 84 78

Fifth Period

75 100 100 69 70 72 60 85 100 40 53 66
81 55 65 75 62 82 58

PROBABILITY 57B

Bird vs. Johnson

	Bird		Johnson	
	Points	Steals	Points	Steals
79–80	21	143	18	187
80–81	21	161	22	127
81–82	23	143	19	208
82–83	24	148	17	176
83–84	24	144	18	150
84–85	29	129	18	113
85–86	26	166	19	113
86–87	28	135	24	138
87–88	30	125	20	114
88–89	*injured*		23	138

PROBABILITY
Problem Set 57

HAMBURGER HEAVEN: Harry Halten, the owner of Hamburger Heaven, wanted to compare the number of hamburgers he sold between 11:00 and 1:00 with the number he sold between 4:00 and 6:00. He collected data for twenty days.

11:00–1:00

37 64 56 60 50 48 47 47 51 66
52 58 62 44 58 50 49 55 57 48

4:00–6:00

58 68 48 50 59 67 78 66 54 49
61 55 60 63 74 64 50 67 59 73

1. Draw an unordered back to back stem and leaf plot and then order the plot.

2. Find the range of the hamburgers sold between
   a. 11:00–1:00   b. 4:00–6:00

3. Find the median number of hamburgers sold between
   a. 11:00–1:00   b. 4:00–6:00

4. Find the upper quartile of the number of hamburgers sold between
   a. 11:00–1:00   b. 4:00–6:00

5. The range of ages of the players on the team was 12 years. The youngest player was 19. What was the age of the oldest player?

6. The median weight of the players was 175 pounds. What does that mean?

7. The upper quartile weight was 200 pounds. What does that mean?

8. What time is the next palindrome after 5:30?

9. The odometer on Jotto's car reads 6,876. What will the mileage be when it next shows a palindrome?

10. Use the reverse and add procedure to create a palindrome from 59.

The number of days that eight students were absent from school is given: 6 5 18 4 4 6 7 2

11. What was the range of the days absent?   12. Find the mean number of days absent.

13. Find the median number of days absent.

14. Is the median or the mean a better description of the data?

SNOW WHITE AND THE SEVEN DWARFS: The average income of the seven dwarfs was $300. Snow White made $5000.

15. What was the total income of the dwarfs?

16. What was the total income of all eight characters?

17. What was the average income of all eight characters?

GUINNESS RECORD®: Charles Walker played 201 games of checkers simultaneously on November 5, 1988 in New Orleans, LA. He won 189 and tied 12.

18. What percent of the games did he win?

PROBABILITY
Lesson Plan 58

QUIZ

Quiz 23

OBJECTIVES

Students will be able to draw a box and whisker plot in order to compare two sets of data.

MATERIALS NEEDED

1. Quiz 23
2. Transparency 58
3. Problem Set 58

CLASS ACTIVITIES

Box and whisker plots are used to compare two or more sets of data. You can compare data gathered from your class or use the data on the NBA All Star teams. Data that you gather from your class might be age in months, pulse rate, height in centimeters, hand span in centimeters, the number of times they can snap their fingers in 30 seconds, etc. You must be careful not to gather any "embarrassing" data. The box and whisker plots would compare boys with girls. To draw a box and whisker plot, you must determine the biggest value, the smallest value, the median, and the upper and lower quartiles. The box is a rectangle showing where the middle 50% of the data lies and the whiskers are lines joining the extreme values to the upper and lower quartiles. A line in the box indicates the median. All of the values are scaled on a number line. The box and whisker plots for the 58–59 and 88–89 NBA All Stars are shown.

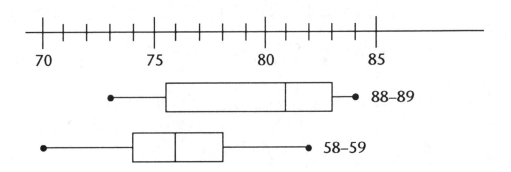

ASSIGNMENT

Problem Set 58

PROBABILITY QUIZ 23

Name _____

The stem and leaf plot gives the ages of people that attended the Simpson Family reunion.

```
4 | 1 2
3 | 1 2 5 5 6 7 7
2 | 3 5 8 8
1 | 0 1
```

1. How old was the oldest person attending? _____

2. What was the range of the ages? _____

3. How many people attended? _____

4. What was the median age? _____

5. What was the upper quartile age? _____

6. Wilbur's watch said 6:57. In how many minutes will it show a palindrome? _____

7. The range of the pulse rates was 25 and the slowest pulse was 49. What was the fastest? _____

8. What does the following mean? The median age of all NHL hockey players is 26. _____

_____

9. Is it easier to calculate the median or range? _____

Basketball Players—Bigger and Bigger

NBA All Star Teams

58–59

Elgin Baylor	6'5"
Bob Pettit	6'9"
Bill Russell	6'10"
Bob Cousy	6'1"
Bill Sharman	6'1"
Paul Arizin	6'4"
Cliff Hagan	6'4"
Dolph Schayes	6'6"
Slater Martin	5'10"
Richie Guerin	6'4"

73–74

John Havlicek	6'5"
Rick Barry	6'7"
K. Abdul Jabbar	7'2"
Walt Frazier	6'4"
Gail Goodrich	6'1"
Elvin Hayes	6'9"
Spencer Haywood	6'8"
Bob McAdoo	6'9'
Dave Bing	6'3"
Norm VanLier	6'2'

88–89

Karl Malone	6'9"
Charles Barkley	6'6"
Akeem Olajowon	7'0"
Earvin Johnson	6'9"
Michael Jordan	6'6"
Tom Chambers	6'10"
Pat Ewing	7'0"
John Stockton	6'1"
Kevin Johnson	6'1"
Chris Mullin	6'7"

PROBABILITY
Problem Set 58

GUINNESS RECORD®: In public life, the fastest talking speed was achieved by President John F. Kennedy in December of 1961. For a short time he spoke at the rate of 300 words per minute.

1. How many words did he speak per second?
2. How many seconds would it have taken him to say the paragraph above?

The box and whiskers plot shows the weights of 36 players on a high school curling team.

3. How much does the heaviest player weigh?
4. What is the range of weights?
5. What is the median weight?
6. What percent of the players weigh more than 175 pounds?
7. How many players weigh more than 175 pounds?
8. What is the upper quartile weight?
9. What percent of the players weigh less than 155 pounds?
10. How many players weigh between 155 and 180 pounds?

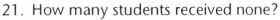

The stem and leaf plot shows the number of pages in each chapter of the book *Pitcher in the Wheat.*

```
3 | 1 4 9
2 | 0 0 4 5 7 8 8 8
1 | 0 3 4 4 5 7 7
```

11. How many chapters are in the book?
12. How many pages in the shortest chapter?
13. What is the range of the chapter lengths?
14. What is the median number of pages?
15. What is the upper quartile?
16. What is the lower quartile?
17. Draw a box and whisker plot of this data.

In a seven-day period, Cassius the Cat killed the following number of mice: 4, 9, 9, 7, 6, 8, 6.

18. What was the mean number of mice killed each day?
19. What was the median number of mice killed each day?
20. What was the range of the number of mice killed each day?

The frequency distribution shows the number of parking tickets that a group of students received.

21. How many students received none?
22. How many received three?
23. What was the range?
24. What was the median number of tickets received?
25. What was the mean number of tickets received?

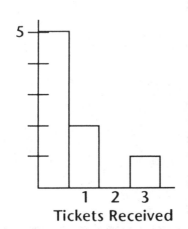

Tickets Received

PROBABILITY
Lesson Plan 59

QUIZ   None

OBJECTIVES   Given two sets of data, students will be able to draw a scatterplot and a trend line if one exists. Students will then be able to state whether there is a positive, negative, or no correlation between the two sets of data.

MATERIALS NEEDED

1. A piece of graph paper for each student
2. Transparency of the graph paper
3. Yard stick
4. Transparencies 59A–C
5. Problem Set 59

CLASS ACTIVITIES   Devise an efficient method for gathering data about the students. Information that could be used includes height, hand span, arm span, forearm length, head circumference, number of brothers and sisters, number of miles they live from school, etc. Some of these quantities are positively correlated, and some have no correlation. It is difficult to come up with quantities that are negatively correlated. Two quantities that probably are positively correlated are hand span and height. Make a table of values:

Student	Height	Hand-span
1	66	$8\frac{1}{2}$
2	72	10

Next construct a scatter plot. It may look like this. This shows that taller people have larger hands.

If one graphed height versus brothers and sisters, there should be no trend line and no correlation.

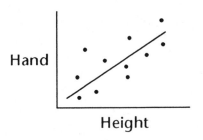

(continued on next page)

PROBABILITY
Lesson Plan 59

There is a difficulty with using student-generated data. The trend line may not be evident. Transparencies 59A and 59B contain data that I have made up. Definite trend lines do exist. There is a positive correlation between years of education and income and a negative correlation between life expectancy and packs of cigarettes smoked per day. A nice "extra credit" project would be for a student to obtain the real data for these events. Transparency 59C should be discussed to show that correlation is not the same as causation. These three transparencies can be used with any of Lessons 59, 60, or 61.

ASSIGNMENT   Problem Set 59

PROBABILITY 59A

Income Versus Years of Education

Ten 25 year olds were surveyed to find their years of education and their annual income.

Years of Education	Annual Income
8	$9,000
10	11,000
12	17,000
12	17,500
13	20,600
14	21,000
16	24,000
16	25,000
18	30,000
19	31,000

Is there a trend?

PROBABILITY 59B

Dr. Nos Moking investigated the relationship between the number of packs of cigarettes 15 people smoked each day and their age at death.

Packs/Day	Age At Death
0	76
7	41
1	56
5	48
2	60
8	40
0	76
3	74
4	52
7	40
1	66
3	52
0	68
1	60
6	48

Is there a trend?

PROBABILITY 59C

Ice Cream and Drownings

The Sealquiz Ice Cream company compared the number of deaths by drowning with the number of ice cream cones sold for a twelve month period.

Month	Drownings	Ice Cream Cones Sold (1,000s)
1	2	4
2	2	5
3	4	6
4	5	8
5	7	11
6	10	16
7	10	18
8	9	16
9	5	7
10	2	6
11	1	3
12	1	3

Is there a trend?

Does eating ice cream cones cause an increase in deaths by drownings?

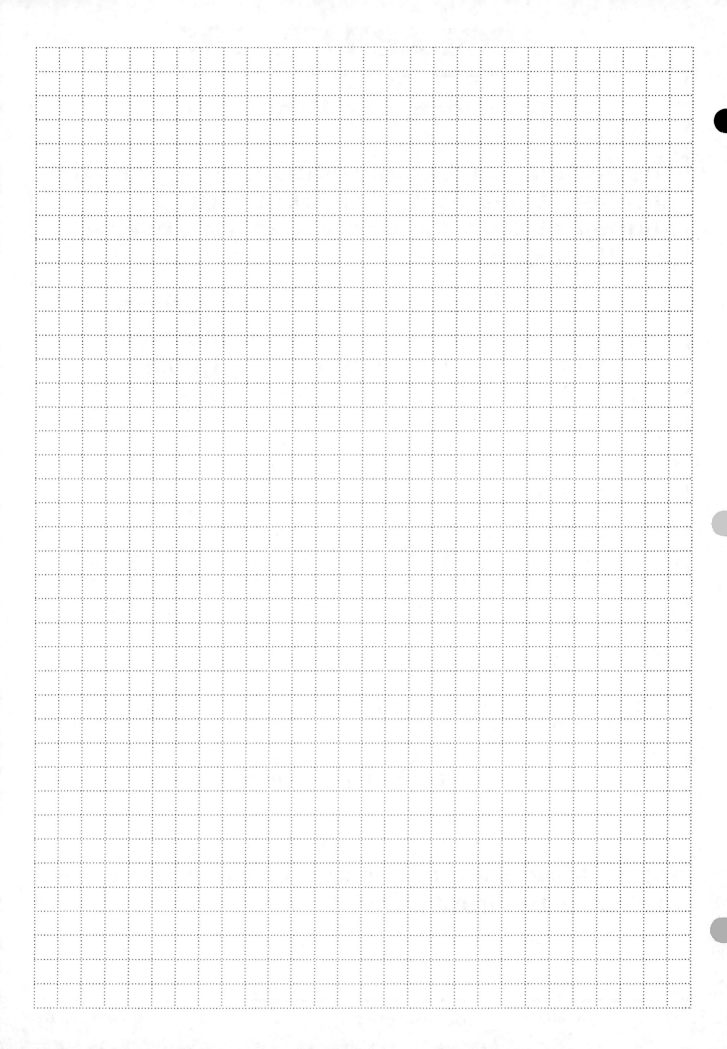

PROBABILITY
Problem Set 59

GUINNESS RECORD®: On June 25, 1988 Reg Morris extinguished 22,888 flame torches in his mouth in two hours at Yule Tree School in West Midlands, England.

1. On the average, how many did he extinguish each minute?

2. On the average, how many did he extinguish each second?

There are 40 ghosts living in Transylvania. The oldest is 60 and the youngest is 38. The median age is 45, the upper quartile is 58, and the lower quartile is 40.

3. Draw a box and whiskers plot of this data. 4. What is the range of ages?

5. How many ghosts are older than 58? 6. How many ghosts are between 40 and 58?

The box and whiskers plot gives information about the hourly wages of 60 workers at Bruno's Bicycle Builders.

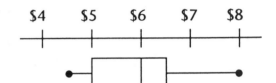

7. What is the highest wage?

8. What is the lowest wage?

9. What is the median wage?

10. What percent of the workers make between $5.00 and $6.50?

11. What percent of the workers make more than $5.00?

12. How many workers make more than $5.00?

In order to determine whether the National League or American League had more powerful hitters, Sam Statistic made a back to back stem and leaf plot comparing home runs hit by National League teams with American League teams for the 1989 season.

National		American
7 1	14	2 5
	13	0 4
8 8 4 3 0	12	2 6 7 7 9
	11	6 7
0	10	1 8
7 5	9	4
9	8	
3	7	

13. How many teams are in the National League?

14. What was the fewest home runs hit by an American League team?

15. Determine the median for:

    a. National League.   b. American League.

16. Determine the upper quartile for the

    a. National League.   b. American League.

17. Determine the lower quartile for the

    a. National League.   b. American League.

18. On one scale, draw a box and whiskers plot for each league.

PROBABILITY
Lesson Plan 60

QUIZ   Quiz 24

OBJECTIVES   This lesson reviews box and whisker plots and scatter plots.

MATERIALS NEEDED   1. Quiz 24
                    2. Transparency 60
                    3. Graph paper
                    4. Problem Set 60

CLASS ACTIVITIES   *Consumer Reports* has many articles with interesting data. The data
                    in Transparency 60 deals with scores that a panel of judges gave to
                    various "walkman" type tape/radio players. The data can be
                    analyzed with the class in a number of ways. Scatter plots can be
                    made to determine if there is any correlation between:
                    i.   price and tape score
                    ii. price and radio score
                    iii. tape score and radio score

                    Additionally, a box and whisker plot can be made for the prices, and
                    the tape scores and radio scores can be compared using box and
                    whisker plots.

ASSIGNMENT   Problem Set 60

PROBABILITY QUIZ 24

Name _____

The box and whiskers plot gives information concerning scores of students on a math test.

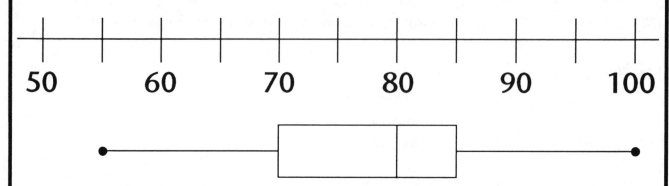

Answer the following. Two are impossible.

1. What is the highest score? _____

2. What is the range of scores? _____

3. What is the median score? _____

4. What is the mean score? _____

5. How many students took the test? _____

6. What is the upper quartile? _____

7. What percent of the students scored above 80?

    _____

8. What percent scored below 70? _____

9. What percent score below 55? _____

10. What percent scored between 70 and 85? _____

PROBABILITY 60

Tape Players and Radios

The October 1991 issue of *Consumer Reports* evaluated 12 tape players.

Model	Price	Score Tape	Radio
Aiwa HS	$49	86	73
Realistic	105	83	81
Sony WM-F2015	33	81	75
Aiwa Radical	89	80	78
Magnavox	56	80	77
GE 3-5474	30	79	71
Sharp	61	79	70
Sanyo	39	78	75
Sears	35	80	77
Sony WM-AF54	69	78	71
GE 3-5477	40	74	78
Emerson	24	73	76

PROBABILITY
Problem Set 60

The following table gives information concerning sandwiches sold at Burger King.

Sandwich	Calories	Fat (g)	Cholesterol (mg)
Whopper	614	36	90
Whopper/Cheese	706	44	115
Double Whopper	844	53	169
Double Whopper/Cheese	935	61	194
Cheeseburger	318	15	50
Cheeseburger Deluxe	390	23	56
Double Cheeseburger	483	27	100
Hamburger	272	11	37
Hamburger Deluxe	344	19	43
Bacon Double Cheeseburger	515	31	105
Bacon Double C'burger Deluxe	592	39	111
BK Broiled Chicken Sandwich	379	18	53
Chicken Sandwich	685	40	82
Ocean Catch Fish Filet	95	25	57

1. Create an ordered stem and leaf plot for the number of grams of fat in each sandwich.

2. What is the range in the amount of fat?   3. What is the median amount of fat?

4. What is the upper quartile of the amount of fat?

5. What is the lower quartile of the amount of fat?

6. Draw a box and whiskers plot of the amount of fat.

7. Draw a scatterplot with calories on the horizontal axis and fat on the vertical axis. If there is a trend line, draw it.

8. Draw a scatterplot with calories on the horizontal axis and cholesterol on the vertical axis. If there is a trend line, draw it.

For each of the following, state whether you think there is a positive correlation, negative correlation, or no correlation.

9. Number of brothers and sisters/weight   10. Hours studied/test score

11. Age/number of words in vocabulary of preschool children

12. Age of car/value of car   13. Age of car/age of driver

14. Age of car/number of miles on odometer   15. Age of driver/speed they drive on highway

GUINNESS RECORD®: The biggest beating handed to a "one arm bandit" (slot machine) was $6,814,823 by Cammie Brewer, 61, at the Club Cal-Neva, Reno, Nevada on February 14, 1988.

16. If she had to pay 30% in taxes, how much money did she pay?

17. How much would remain after taxes?

18. If she planned to spend the remainder in 15 years, how much would she need to spend

    a.   per year?   b.   per day?

PROBABILITY
Lesson Plan 61

QUIZ   None

OBJECTIVES   This is a review lesson.

MATERIALS NEEDED   1. Transparency 61
                    2. Problem Set 61

CLASS ACTIVITIES   Using data from Transparency 61, various stem and leaf, box and whisker, and scatterplots can be drawn. Suggested exercises to work through with the students are:

i.   frequency distribution of cost
ii. frequency distribution of fat
iii. stem and leaf plot of scores
iv. box and whisker plot of scores
v.   scatterplot of score versus cost
vi. scatterplot of fat versus calories

ASSIGNMENT   Problem Set 61

PROBABILITY 61

Yogurt

Brand	Type	Cost	Score	Calories	Fat
Columbo	W	9	90	150	7
Dannon	L	7	85	150	4
Mountain High	W	7	85	200	9
Yoplait	L	12	80	173	4
A & P	L	5	74	150	4
Columbo	N	6	73	110	0
Yoplait	N	6	70	120	0
Kroger	L	7	65	140	4
Axelrod	L	5	64	140	2
Breyers	L	6	58	140	3
Dannon	N	7	52	130	0
Albertson's	L	4	40	140	5
Lucerne	L	5	30	160	5
Weight Watcher	N	8	12	90	1

Consumer Reports, May 1991

Cost: per ounce
Type: W – at least 3% Milkfat
L – between 1/2% and 2% Milkfat
N – made from skim milk
Fat: grams

PROBABILITY
Problem Set 61

BROWLER the BOWLER: The following are the scores for nine games of bowling rolled by Browler: 146, 183, 141, 161, 176, 177, 152, 174, 177.

1. Determine the mean score.
2. Construct an ordered stem and leaf plot.
3. Determine the median score.
4. Determine the upper quartile.
5. Determine the lower quartile.
6. Draw a box and whiskers plot.

STUDIOUS STUDENTS: The frequency distribution gives the number of hours a group of students studied for the driver's test.

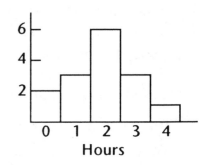

7. What was the greatest number of hours studied?
8. How many students studied 2 hours?
9. How many students took the driver's test?
10. Determine the mean number of hours studied.
11. Determine the median number of hours studied.

STUDIOUS STUDENTS (part two): The scatter plot shows the scores received by a second group of students on the driver"s test.

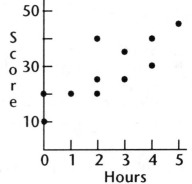

12. How many students took the test?
13. What was the highest score?
14. What was the lowest score?
15. What was the greatest number of hours studied?
16. Is there a positive correlation between hours studied and score?
17. The average yearly income of six teenagers was $560. What was their total income?
18. The median income of six teenagers was $520. How many made more than $520?
19. The range of incomes of six teenagers was $70. The most that anyone made was $650. What was the least that any on made?
20. Six teenagers made a total of $3402. What was their average income?

The boss made $1945 and the five teenagers averaged $456.

21. What was the total amount of money made by the six people?
22. What was the average salary of the six workers?

STUDIOUS STUDENTS (part three): The box and whiskers plot gives information about math SAT scores for a group of students.

23. What was the highest score?
24. What was the median score?
25. What percent of the students scored above 450?
26. What percent of the students scored between 450 and 650?

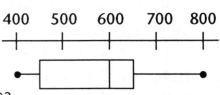

PROBABILITY
Test 6

Do not write on this test. Do all work on scratch paper and put the answers in the appropriate space on the answer sheet.

GUINNESS RECORD®: Brian Keene and James Wright drove their Chevrolet Blazer 9,031 miles in 37 days (August 1–September 7, 1984) in *reverse* through 15 states and Canada.

1. If they each drove the same distance, how far did each drive?
2. How many miles did they average each day? (whole number)
3. If they drove for sixteen hours each day, how many miles did they average each hour? (whole number)

Hortence the Hoopster: In seven games, Hortence scored 14, 16, 11, 3, 11, 15, and 15 points.

4. What was the range of the points that she scored?
5. What was the mean number of points that she scored? (nearest tenth)
6. What was the median number of points that she scored?
7. Does the mean or median give a truer representation of her average?

Horatio the Hoopster: Horatio averaged 11 points in six games. His range was nine points and the most points he scored was 15.

8. What was the total number of points he scored in the six games?
9. What was the fewest points that he scored?

Horatio scored 20 points in the seventh game.

10. What was the total number of points he scored in seven games?
11. What was his average (mean) points scored in seven games? (nearest tenth)

Peperus the Pianist: The stem and leaf plot gives information about the number of hours that Peperus spent practicing the piano each week.

```
4 | 3 4 7
3 | 0 2 5 5 8
2 | 5 8
1 | 7 9 9
```

12. How many weeks did Peperus practice?
13. What's the greatest number of hours?
14. What was the range of his practice time?
15. What was the median?
16. What was the upper quartile?
17. What was the lower quartile?

Solyam's Soda Shoppe: The box and whiskers plot gives information about the number of sodas sold at Solyam's for the 28 days of February.

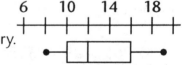

18. What was the greatest number sold?
19. What was the range of the number sold?
20. What was the median number sold?
21. What percent of days were more than 16 sold?
22. How many days were more than 16 sold?
23. How many days were between 10 and 16 soda's sold?

PROBABILITY
Test 6

Daughter Data: The frequency distribution gives data concerning the number of daughters in teachers' families.

24. How many teachers had no daughters?

25. What was the greatest number of daughters any teacher had?

26. What was the range?

27. How many families are represented?

28. What was the median number of daughters?

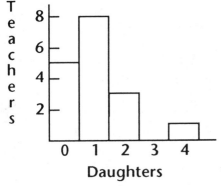

True–False

29. If all numbers in a set of data are 7, the range is 7.

30. If all numbers in a set of data are 7, the mean is 7.

31. If all numbers in a set of data are 7, the median is 7.

32. There is probably a positive correlation between one's shoe size and height.

33. There is probably no correlation between one's Zip Code and shoe size.

34. There is probably a negative correlation between an adult's age and the time it takes him to run a mile.

35. A scatter plot can be used to determine correlation.

36. A stem and leaf plot lets you determine how many pieces of data have been collected.

37. A box and whiskers plot allows you to determine the mean.

38. The median of a set of data is always larger than the mean.

39. It is easier to calculate the range than the mean.

Solyam's Soda Shoppe Revisited: The scatterplot gives information concerning the number of sodas sold and the temperature over a ten day period.

40. What was the greatest number of sodas sold?

41. What was the fewest number of sodas sold?

42. How many sodas were sold when the temperature was 60°?

43. Is the correlation positive?

44. The odometer on the car showed 4644. How many miles does Dringa need to drive before it shows a palindrome?

45. Sasquach set her digital alarm clock for a palindromic time so that she would get up shortly before 7:00 am. What time did the alarm go off?

46. What palindrome do you end up with if you use the reverse and add procedure starting with 84?

47. Write one palindromic word.

PROBABILITY

Test 6 Answer Sheet

Name _____

1. _____   17. _____   33. _____

2. _____   18. _____   34. _____

3. _____   19. _____   35. _____

4. _____   20. _____   36. _____

5. _____   21. _____   37. _____

6. _____   22. _____   38. _____

7. _____   23. _____   39. _____

8. _____   24. _____   40. _____

9. _____   25. _____   41. _____

10. _____   26. _____   42. _____

11. _____   27. _____   43. _____

12. _____   28. _____   44. _____

13. _____   29. _____   45. _____

14. _____   30. _____   46. _____

15. _____   31. _____   47. _____

16. _____   32. _____

PROBABILITY
Answers to Problem Sets

Problem Set 1

1. 34%	2. 56%	3. 50%	4. 12%	5. 75%	6. 85%
7. 45%	8. 8%	9. 146%	10. 23%	11. 22%	12. 108%
13. 3/5	14. 1/4	15. 1	16. 9/20	17. 2	18. 2/5
19. 3/2	20. 12/25	21. 11/50	22. 1/10	23. 8/5	24. 3/4

25. to 30. Answers vary.   31. ≈55.56 mi. 32. ≈5.5 hrs.   *BRAIN BUSTER:*

Problem Set 2

1. 6%	2. 50%	3. 52%	4. 20%	5. 85%	6. 75%
7. 110%	8. 9%	9. 246%	10. 27%	11. 56%	12. 46%
13. 42%	14. 21%	15. 111%	16. 83%	17. 19%	18. 5%
19. 7/20	20. 16/25	21. 6/5	22. 1	23. 1/4	24. 2/25

25.
Dimes	Nickels	Pennies
2	0	3
1	2	3
1	1	8
1	0	13
0	4	3
0	3	8
0	2	13
0	1	18
0	0	23

26. 5   27. 23

28. 1 dime, 2 nickels, 3 pennies

29. no   30.

Five	Ten	Twenty	Total
3	0	0	15
2	1	0	20
2	0	1	30
1	2	0	25
1	1	1	35
1	0	2	45
0	3	0	30
0	2	1	40
0	1	2	50
0	0	3	60

31. 60

32. 15   33. 9   34. 2

Problem Set 3

1. 74%	2. 44%	3. 44%	4. 80%	5. 50%	6. 190%
7. 7/20	8. 16/25	9. 6/5	10. 1	11. 3/4	12. 2/25
13. 92%	14. 58%	15. 18%	16. 23%	17. 142%	18. 167%

19. 1/5 = 20%   20. 0/5 = 0%   21. 3/5 = 60%   22. 2/5 = 40%

23. 5/5 = 100%   24. 3/5 = 60%   25. 2/5 = 40%   26. 1/5 =20%

27.
Long	Fg	Ft
3	0	0
2	1	1
2	0	3
1	3	0
1	2	2
1	1	4
1	0	6
0	4	1
0	3	3
0	2	5
0	1	7
0	0	9

28. two ways   29. five ways

BRAIN BUSTER: There are 15 rectangles.

PROBABILITY
Answers to Problem Sets

Problem Set 4

1. 75% 2. 102% 3. 17% 4. 61% 5. 16% 6. 50%

7. 16% 8. 54% 9. 78% 10. 58% 11. 21% 12. 22%

13. 55% 14. 85% 15. 3/4 16. 997/1000 17. 20/43 18. 1/5 = 20%

19. 1/5 = 20% 20. 0/5 = 0% 21. 5/5 = 100% 22. 3/5 = 60%

23. 4/5 = 80% 24.

TD	FG	EP
3	0	3
3	1	0
2	3	0
1	5	0
0	7	0

25. one way

26.

TD	FG	EP
5	0	0
4	2	0
4	1	3
3	4	0
3	3	3
2	6	0
1	8	0

27. five ways

28. ten, eleven, or twelve times

Problem Set 5

1. 1/10 = 10% 2. 0/10 = 0% 3. 10/10 = 100% 4. 2/10 = 20%

5. 7/10 = 70% 6. 0/10 = 0% 7. 0/10 = 0% 8. 6/10 = 60%

9. 9/10 = 90% 10. 3/10 = 30% 11. 24% 12. 1/5

13. 1/7 14. 92% 15. 75% 16. 115% 17. 33% 18. 22%

19. 123% 20. 1/2 21. 3/4 22. 1/10 23. 1 24. 92%

25. 11% 26. 36% 27. 21% *BRAIN BUSTER:* The starting number is 3.

Problem Set 6

1. 1/6 2. 5/6 3. 1/2 4. 1/2 5. 1/4 6. 1/8

7. 1/16 8. 15/16 9. 1/2 10. 3/4 11. 1/4 12. 1/2

13. 1/8 14. 7/8 15. 3/8 16. 1/4 17. 1/8 18. 7/8

19. 3/4 20. 3/9 = 33% 21. 6/9 = 67% 22. 1/9 = 11%

23. 0/9 = 0% 24. 4/9 = 44% 25. 2/9 = 22% 26. 5/9 = 55%

27. 5/9 = 55% 28. 4/9 = 44% 29. 5/9 = 55

Problem Set 7

1. 1/3 2. 2/3 3. 2/3 4. 1 5. 3/4 6. 1/4

7. 1/8 8. 7/8 9. 1/4 10. 1/6 11. 5/6 12. 1/3

13. 1/2 14. 1/4 15. 1/8 16. 7/8 17. 1/2 18. 1/4

19. 1/8 20. 1/4 21. 1/8 22. 1/4 23. 12/25 = 48%

24. 5/25 = 20% 25. 2/25 = 8% 26. 9/25 = 36% 27. 4/25 = 16%

28. 5/25 = 20% 29.

1¢	5¢	10¢
20	0	0
15	1	0
10	2	0
10	0	1
5	3	0
5	1	1
0	4	0
0	2	1
0	0	2

30. 2 coins 31. 20 coins

PROBABILITY
Answers to Problem Sets

Problem Set 8

1. 1/2   2. 1/4   3. 1/8   4. 7/8   5. 3/4   6. 3/8
7. 1/4   8. 3/4   9. 1/8   10. 7/8   11. 1/3   12. 1/3
13. 1/2   14. 1/6   15. 2/3   16. 0   17. BATMAN   18. 98%
19. 49,997/50,000   20. ≈155 words   21. ≈3.2 min.

BRAIN BUSTER: Freeda is 21.

Problem Set 9

1. 50   2. 19   3. 19/50 = 38%   4. 31/50 = 62%
5. 23/50 = 46%   6. 35/50 = 70%   7. 7/50 = 14%   8. 84
9. 78   10. 200   11. 122/200 = 61%   12. 84/200 = 42%
13. 116/200 = 58%   14. 154/200 = 77%   15. 46/200 = 23%   16. 1/6
17. 1/2   18. 1/4   19. 1/3   20. 5/6   21. 1/4   22. 1/4
23. 1/8   24. 1/2   25. 7/8

26.
Y	T
v	p
v	m
v	f
s	p
s	m
s	f
c	p
c	m
c	f

Problem Set 10

1. 45%   2. 34%   3. 12%   4. 18%
5. 70%   6. 5%   7. 1/2   8. 3/4
9. 11/10   10. 11/25   11. 1/50   12. 16/25
13. 71%   14. 89%   15. 88%   16. 94%
17. 63%   18. 36%   19. 96%   20. 7/20   21. 1/3   22. 2/3
23. 0   24. 1/2   25. 2/3   26. 1/6   27. 1   28. 1/4
29. 1/4   30. 1/8   31. 3/4   32. 1/6   33. 2/3   34. 5/6
35. 1/3   36. 1/8   37. 3/4   38. 7/8   39. 7/8   40. 100
41. 58   42. 17/25   43. 29/50   44. 81/100   45. 9/20

Problem Set 11

1. 7   2. 2 and 12   3. 9   4. 1/18   5. 5/18   6. 1/12
7. 1/9   8. 2/3   9. 7/36   10. 1/4   11. 45%   12. 47%
13. 3%   14. 25%   15. 80%   16. 50%   17. 1%   18. 114%
19. 25%   20. 68%   21. 75%   22. 50%   23. 33%   24. 62%
25. 56%   26. 4%

27.
x	1	2	3
1	1	2	3
2	2	4	6
3	3	6	9

28. 1/9   29. 0
30. 1   31. 5/9
32. 4/9   33. 1/3
34. 8/9   *BRAIN BUSTER:*

PROBABILITY
Answers to Problem Sets

Problem Set 12

1. 68%   2. 50%   3. 44%   4. 89%   5. 33%   6. 37%
7. 90%   8. 70%   9. 112%   10. 28%   11. 64%   12. 111%
13. 95%   14. 1%   15. 6%

16.

+	1	2	3	4	5	6
1	2	3	4	5	6	7
2	3	4	5	6	7	8
3	4	5	6	7	8	9

17. 4, 5, 6 or 7   18. 1/18 or 6%   19. 6/18 or 33%
20. 12/18 or 67%   21. 9/18 or 50%   22. 7/18 or 39%
23. 14/18 or 78%   24. 1/6   25. 5/6
26. 1/3   27. 5/6   28. 1/3   29. 2/3   30. 5/6
31. 5/6   32. 28.5 feet   33. 320 pounds

Problem Set 14

1. 34%   2. 83%   3. 7%   4. 88%   5. 12%   6. 68%
7. 52%   8. 39%   9. 9%   10. 1%   11. 6481   12. 99%

13.

+	1	2	3	5
1	2	3	4	6
2	3	4	5	7
3	4	5	6	8
5	6	7	8	10

14. 4 or 6   15. 6%   16. 0%   17. 38%
18. 31%   19. 75%   20. 19%   21. 1/8
22. 1/4   23. 7/8   24. 1/4   25. 3/8
26. 7/8   27. 97%

Problem Set 15

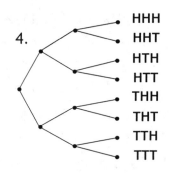

5. 26%   6. 35%   7. 64%   8. 90%   9. 14%   10. 105%
11. 27%   12. 105%   13. 44%   14. 20%   15. 60%   16. 100%
17. 19%   18. 279   19. 81%   20. 87%   21. 45   22. 13%

Problem Set 16

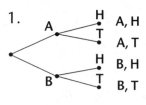

2. 1/8   3. 1/8   4. 1/8   5. 1/2
6. 1/4   7. 3/4   8. 1/4
9. 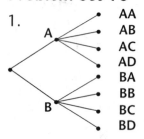   10. $40   11. $2

PROBABILITY
Answers to Problem Sets

12. 1/9   13. 1/9   14. 2/9   15. 8/9   16. 1/9   17. 1
18. 62%   19. 111%   20. 85%   21. 80%   22. 90%   23. 9%
24. 115%   25. 13%   26. 1%   27. 50%   28. 10%   29. 0%
30. 71%   *BRAIN BUSTER:* 18 nickels, 1 dime

Problem Set 18

1. Ivan has 13 cents.
2.

Dimes	Nickels	Pennies
1	0	3
0	2	3
0	1	8
0	0	13

3. Abigail is 44.
4. 74%   5. 44%
6. 100%   7. 50%
8. 90%   9. 35%
10.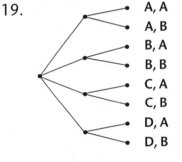
11. 1/9   12. 2/9
13. 1/3   14. 1/9   15. 2/3   16. 4/9
17. 16   18. 45 miles   19. 5 miles

Problem Set 19

1. Delwyn's IQ is 85.
2. Jessica did 14 fingersnaps.
3. Delwyn made $16 the first day.   4. 74%
5. 12%   6. 96%
7. 100%   8. 88%   9. $1,495   10. 12%
11.

	4	**5**	**6**
4	9	10	11
5	10	11	12
6	11	12	13

12. 13   13. 11   14. 4/9   15. 1/3
16. 2/9   17. 5/9   18.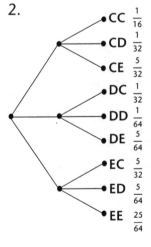

H — H → H, H
H — T → H, T
T — H → T, H
T — T → T, T

19.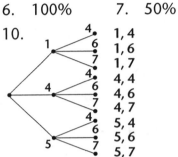

A, A
A, B
B, A
B, B
C, A
C, B
D, A
D, B

BRAIN BUSTER: 8

Problem Set 21

1.
AA $\frac{1}{16}$
AB $\frac{3}{16}$
BA $\frac{3}{16}$
BB $\frac{9}{16}$

2.
CC $\frac{1}{16}$
CD $\frac{1}{32}$
CE $\frac{5}{32}$
DC $\frac{1}{32}$
DD $\frac{1}{64}$
DE $\frac{5}{64}$
EC $\frac{5}{32}$
ED $\frac{5}{64}$
EE $\frac{25}{64}$

3.
FF $\frac{49}{64}$
FG $\frac{7}{64}$
GF $\frac{7}{64}$
GG $\frac{1}{64}$

4. Tree as in Problem 2.

HH $\frac{1}{4}$
HI $\frac{1}{12}$
HJ $\frac{1}{6}$
IH $\frac{1}{12}$
II $\frac{1}{36}$
IJ $\frac{1}{18}$
JH $\frac{1}{6}$
JI $\frac{1}{18}$
JJ $\frac{1}{9}$

5. 50%   6. 26%   7. 39%   8. .56   9. .38
10. 1.0   11. .287   12. .06   13. 1.56   14. 20
15. 49%   16. 51%   17. 7/12   18. 1/3   19. 1/12
20. 1/2   21. 1/4   22. 1/4   23. 1/3   24. 1/6
25. 1/2   26. 1/12   27. 3/4   28. 1/6

PROBABILITY
Answers to Problem Sets

Problem Set 22

1.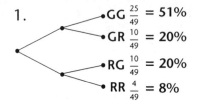
GG $\frac{25}{49}$ = 51%
GR $\frac{10}{49}$ = 20%
RG $\frac{10}{49}$ = 20%
RR $\frac{4}{49}$ = 8%

2. G,G   3. R,R   4. 29/49   5. 20/49
6. 4 times   7. 8 times   8. 40 times   9. 8 times
10. 4 times   11. 32 times   12. 20 times   13. .37
14. .4   15. .04   16. 1.34   17. .469
18. .018

19.
AAA $\frac{27}{64}$ = 42%
AAB $\frac{9}{64}$ = 14%
ABA $\frac{9}{64}$ = 14%
ABB $\frac{3}{64}$ = 5%
BAA $\frac{9}{64}$ = 14%
BAB $\frac{3}{64}$ = 5%
BBA $\frac{3}{64}$ = 5%
BBB $\frac{1}{64}$ = 2%

20. Tree as in Problem 19.
AAA $\frac{8}{125}$ = 6%
AAB $\frac{12}{125}$ = 10%
ABA $\frac{12}{125}$ = 10%
ABB $\frac{18}{125}$ = 14%
BAA $\frac{12}{125}$ = 10%
BAB $\frac{18}{125}$ = 14%
BBA $\frac{18}{125}$ = 14%
BBB $\frac{27}{125}$ = 22%

21. 58%
22. 21%
23. 0%
24. 24 miles
25. 96 laps

Problem Set 23

1. 64   2. 32   3. 14   4. 0
5. 564   6. 3.5   7. .45   8. .428   9. .09   10. .006
11. 1.74   12. 1   13. $586   14. 90%   15. 10%   16. 20%
17. 80%   18. 520   19. 4 times   20. 8 times   21. 36 times   22. 82 times
23. 41 times   24. 656 times   25. 779 times   26. Probabilities are:
27. all tails
28. 36/125 or 29%   29. 35/125 or 28%
30. 27 times   31. 54 times   32. 22 times   33. 88 times
34. G = 3, M = 9   35. A = 5, L = 0, H = 1

TTT 22%
TTH 14%
THT 14%
THH 10%
HTT 14%
HTH 10%
HHT 10%
HHH 6%

Problem Set 24

1. 230 people   2. 6 questions   3. 34 pounds   4. 24 harmonicas   5. .56
6. .6   7. .06   8. .457   9. 1.34   10. .034   11. 35%
12. 65%   13. 302   14. 64 times   15. 32 times   16. 512 times   17. 7/10
18.
RR 49%
RB 21%
BR 21%
BB 9%

19. R,R   20. 58%   21. 42%   22. 49 times
23. 25 times   24. 147 times   25. 490 times   26. G=2, O=5
27. O=4, N=3, E=2, T=8, W=6 is one solution.
BRAIN BUSTER: four

Problem Set 25

1. 48 people   2. 240 people   3. 120 people   4. 24%
5. 76%   6. 352 people   7. I=3, C=2, A=1, N=5 is one solution
8. L = 1, O = 4, K = 3, A = 7   9. 8 times   10. 2 times   11. 40 times   12. 42 times
13. See Problem Set 22, problem 19. Reverse A and B.   14. 2%   15. 44%
16. 42%   17. 2 times   18. 8 times   19. 1 time   20. 2 times

PROBABILITY
Answers to Problem Sets

21.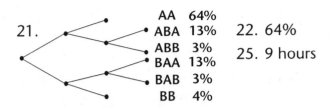
AA   64%
ABA 13%
ABB 3%
BAA 13%
BAB 3%
BB   4%

22. 64%   23. 26%   24. 538 minutes

25. 9 hours

Problem Set 27

1. $3.00   2. $0.25   3. $23.00   4. $447.50   5. $1839.00   6. $0.04

7. 240   8. 48   9. 40%   10. 22%   11.

12. 39%   13. 47%   14. 14%   15. 14 times

16. 7 times   17. 56 times   18. 63 times   19. 9/10

20. 1/10   21.

22.

23. 5 quarters,
    13 nickels

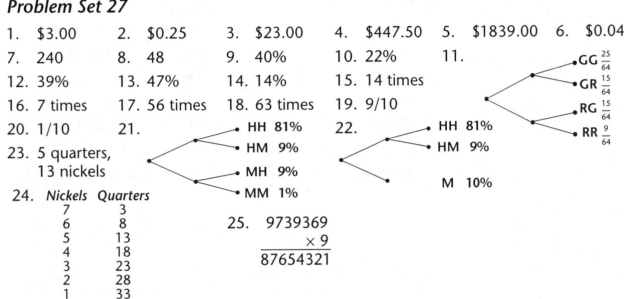

24.
Nickels	Quarters
7	3
6	8
5	13
4	18
3	23
2	28
1	33

25.   9739369
      $\times$ 9
      ─────────
      87654321

Problem Set 28

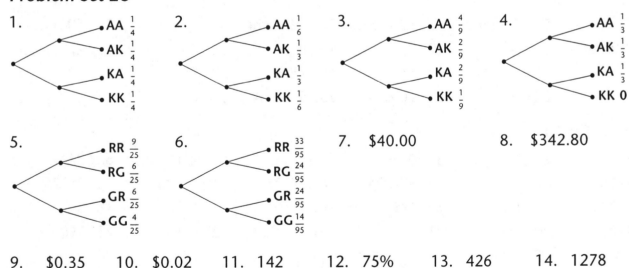

1. AA $\frac{1}{4}$ AK $\frac{1}{4}$ KA $\frac{1}{4}$ KK $\frac{1}{4}$

2. AA $\frac{1}{6}$ AK $\frac{1}{3}$ KA $\frac{1}{3}$ KK $\frac{1}{6}$

3. AA $\frac{4}{9}$ AK $\frac{2}{9}$ KA $\frac{2}{9}$ KK $\frac{1}{9}$

4. AA $\frac{1}{3}$ AK $\frac{1}{3}$ KA $\frac{1}{3}$ KK 0

5. RR $\frac{9}{25}$ RG $\frac{6}{25}$ GR $\frac{6}{25}$ GG $\frac{4}{25}$

6. RR $\frac{33}{95}$ RG $\frac{24}{95}$ GR $\frac{24}{95}$ GG $\frac{14}{95}$

7. $40.00   8. $342.80

9. $0.35   10. $0.02   11. 142   12. 75%   13. 426   14. 1278

15. 73%   16. 27%   17. $1.50   18. $4.20   19. $10.20   20. $19.50

21. .45   22. .7   23. .07   24. .059   25. 74%   26. 2303 species

BRAIN BUSTER:

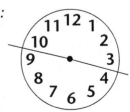

PROBABILITY
Answers to Problem Sets

Problem Set 29

1.
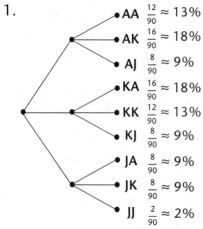
- AA $\frac{12}{90} \approx 13\%$
- AK $\frac{16}{90} \approx 18\%$
- AJ $\frac{8}{90} \approx 9\%$
- KA $\frac{16}{90} \approx 18\%$
- KK $\frac{12}{90} \approx 13\%$
- KJ $\frac{8}{90} \approx 9\%$
- JA $\frac{8}{90} \approx 9\%$
- JK $\frac{8}{90} \approx 9\%$
- JJ $\frac{2}{90} \approx 2\%$

2. 13% 3. 2% 4. 28% 5. 72%
6. 38% 7. 2 times 8. 6 times 9. 1 time
10. 9 times 11.
- RRR $\frac{210}{720} \approx 29\%$
- RRG $\frac{126}{720} \approx 18\%$
- RGR $\frac{126}{720} \approx 18\%$
- RGG $\frac{42}{720} \approx 6\%$
- GRR $\frac{126}{720} \approx 18\%$
- GRG $\frac{42}{720} \approx 6\%$
- GGR $\frac{42}{720} \approx 6\%$
- GGG $\frac{6}{720} \approx 1\%$

12. 29%
13. 1% 14. 18%
15. 30% 16. 70%
17. 1150 18. 75%
19. 3450 20. 44%
21. 474 22. 56%
23. $6.50 24. $136.50
25. $3.50 26. $2.70

27. No, she needs $20.70. 28. P = 4, A = 5, T = 9 29. A = 4, B = 5, P = 7, Q = 6

Problem Set 30

1.
- BB $\frac{25}{49}$
- BW $\frac{10}{49}$
- WB $\frac{10}{49}$
- WW $\frac{4}{49}$

2.
- BB $\frac{20}{42}$
- BW $\frac{10}{42}$
- WB $\frac{10}{42}$
- WW $\frac{2}{42}$

3.
- H $\frac{5}{6}$
- MM $\frac{5}{36}$
- WW $\frac{1}{36}$

4. 42 times 5. 21 times 6. 126 times 7. 189 times 8. 5 times 9. 50 times
10. 3 times 11. 64 12. 3 13. 0 14. 54 15. 27
16. 81 17. 21 18. 11 19. 78% 20. 18 21. 22%
22. 4 23. 36 24. 90% 25. S = 6, A = 2, K = 4

Problem Set 31

1. 7/15 2. 1/15 3. 0 4. 2/15 5. 2/15 6. 2/3
7. 2/15 8. 1 9. 49/225 10. 7/225 11. 2/225 12. 7/225
13. 8/225 14. 1/225 15. 1/5 16. 1/30 17. 1/30 18. 0
19. 1/70 20. 4/15 21.
- PP $\frac{81}{100}$
- PA $\frac{9}{100}$
- AP $\frac{9}{100}$
- AA $\frac{1}{100}$

22. 81% 23. 1%
24. 18% 25. 86% 26. 77% 27. 23%
28. 24,309 29. 1,013 30. 17 31. 71%

Problem Set 32

1. 3/5 2. 1 3. 1/5 4. 3/5 5. 9/25 6. 1/25
7. 2/25 8. 1/25 9. 3/10 10. 1/20 11. 1/10 12. 0
13. 1/10 14. 1/10 15. 24, 29, 34 16. 81, 243, 729 17. 33, 41, 50
18. 32, 26, 20 19. 4, 2, 1 20. 100 21. 42% 22. 58% 23. 71%

PROBABILITY
Answers to Problem Sets

24. 29%   25. 25%   26. 6%   27. 25 times   28. 12 times   29. 74 times

30. 20 times   *BRAIN BUSTER:* Al spent $82.00.

Problem Set 33

1. 4, 1/5, 5
2. 35, 43, 51
3. 162, 486, 1458
4. 19, 25, 24
5. 55555, 666666, 7777777
6. 5, 2.5, 1.25
7. 11
8. 6
9. 3
10. 9
11. 37
12. 43
13. 1/6 = 17%
14. 1/2 = 50%
15. 1/4 = 25%
16. 1/2 = 50%
17. 1/12 = 8%
18. 7/12 = 58%
19. 34.7%
20. 86.3%
21. 0.1%
22. 75.0%
23. 1/46 = 2.2%
24. 5/92 = 5.4%
25. 11/46 = 23.9%
26. 2/69 = 2.9%
27. 5/552 = 0.9%
28. 1/46 = 2.2%
29. 72%
30. 28%
31. 18
32. 62%
33. 38%
34. 119,000 pounds
35. 35 pounds

Problem Set 34

1. 25, 36, 49
2. 12, 13, 15
3. 30, 35, 40
4. h, j, k
5. x, d, w
6. i, h, j
7. 76.5%
8. 65.0%
9. 13.1%
10. .5%
11. 12.8%
12. 187.5%
13. 4
14. 4
15. 2/17 = 11.8%
16. 1/17 = 5.9%
17. 12/425 = 2.8%
18. 11/850 = 1.3%
19. $47
20. $517
21. $3.75
22. 28/45 = 62.2%
23. 1/45 = 2.2%
24. 8/45 = 17.8%
25. 8/45 = 17.8%
26. 76%
27. 24%
28. 111
BRAIN BUSTER:

Problem Set 35

1. 3
2. 2
3. 0, 5, 0
4. f, x, g
5. g, 9, i
6. 3/4, 7/8, 1
7. 6.7%
8. 6.7%
9. 20%
10. 20 times
11. 10 times
12. 80 times
13. $2.40
14. $7.20
15. $12
16. $24
17. 1 5 10 10 5 1
18. 1/32
19. 10/32
20. 10/32
21. 1/32
22. 16/32
23. 2/32
24. 1 7 21 35 35 21 7 1
25. 1/128
26. 21/128
27. 21/128
28. 8/128
29. 64/128
30. over 5 days
31. 17 bars

Problem Set 37

1a. 6   b. 15   c. 10   2a. 24   b. 12   c. 17   3a. 4   b. 6   c. 5
4a. 19   b. 38   c. 25   5. 1 7 21 35 35 21 7 1   6. 1/128 = 1%
7. 1/128 = 1%   8. 70/128 = 55%   9. 21/128 = 16%   10. 0/128 = 0%
11. 16 times   12. 115 times   13. 7%   14. 10%   15. 0%   16. 8%
17. 8%   18. 3%   19. 21, 25, 29   20. 8, e, 10   21. h, 9, j
22. 1250, 6250, 31250   23. 3/5, 7/10, 4/5   24. 74%   25. 26%   26. 1222

PROBABILITY
Answers to Problem Sets

Problem Set 38

1.

5G	4G	3G	2G	1G	0G
1	5	10	10	5	1

2. 1/32 = 3%   3. 10/32 = 31%
4. 10/32 = 31%   5. 16/32 = 50%

6. 3 times   7. 18 times   8. 2 times   9. 14 times   10. 43   11. 22
12. 32   13. 6   14. 67   15. 36   16. 5   17. 421
18. 23.7%   19. 5.3%   20. 11.8%   21. 3.2%   22. 0.8%   23. 3 times
24. 15 times   25. 15 times   26. GLIDE   27. 27 schools   28. 7 days

BRAIN BUSTER: ninth day

Problem Set 39

1. 230   2. 46   3. 170   4. 25%   5. 7%   6. 22, 29, 37
7. 64, 128, 256   8. h, g, j   9. 1/15 = 7%   10. 1/30 = 3%   11. 0 = 0%
12. 1/30 = 3%   13. 4/15 = 27%   14. 8 ways   15. 16 ways   16. 13 times
17. 7 times   18. 68 times   19. 135 times   20.

5M	4M	3M	2M	1M	0M
1	5	10	10	5	1

21. 1/32 = 3.1%   22. 1/16 = 6.3%   23. 5/8 = 62.5%
24. 1/2 = 50%   25. GIGGLE

Problem Set 40

1. LEO   2. 15 ways   3. 63 ways   4. 127 ways   5. 4/7, 13/23, 14/25
6. 16p, 19s, 22v   7.

4H	3H	2H	1H	0H
1	4	6	4	1

8. 1/16 = 6%   9. 1/4 = 25%
10. 1/8 = 13%   11. 3/8 = 38%   12. 6 times
13. 3 times   14. 22 times   15. 9/100 = 9%   16. 6/100 = 6%
17. 16/100 = 16%   18. 25/100 = 25%   19. 1/15 = 6.7%   20. 1/15 = 6.7%
21. 8/45 = 17.8%   22. 2/9 = 22.2%   23. 350   24. 175
25. 322   26. 609

Problem Set 41

1. $0.50   2. -$0.50   3. $4   4. $8   5. -$4   6. $30
7. $51   8. $1026   9. $8   10. $4   11. 94   12. 47
13. 658   14. 431   15. 2244   16. 43%   17. 57%   18. 46
19. 90%   20. 414   21. 1,556 each hour   22. 26 each minute

Problem Set 42

1. $5   2. -$0.31   3. $6.25   4. $14   5. $40   6. $1786
7. $35   8. $234.50   9. $3,370.50   10. $3   11. $0.75   12. 125
13. 190   14. 475   15. 30%   16. 10, 3, 8   17. 4, .5, 3   18. BOBSLED
                                      5, 7, 9   1.5, 2.5, 3.5
                                      6, 11, 4   2, 4.5, 1

PROBABILITY
Answers to Problem Sets

Problem Set 43

1. -$0.50   2. 1/36   3. 25/36   4. even, a fair game   5. $2 win
6. $0.50 loss 7. $2.50 loss 8. $0   9. $90   10. $48   11. $12.90
12. hose   13. 65%   14. 39   15. 0%   16. 7, 0, 5   17. 4, 1.5, 2
                                              2, 4, 6   .5, 2.5, 4.5
                                              3, 8, 1   3, 3.5, 1

18. 18, 4, 5, 15   BRAIN BUSTER: 15 pennies, 15 nickels, 1 dime
    7, 13, 12, 10
    11, 9, 8, 14
    6, 16, 17, 3

Problem Set 44

1. 13% increase   2. 25% increase   3. 8% increase   4. 20% increase
5. 100% increase   6. 10% decrease   7. $70.00   8. $0.60
9. even   10. $26.67   11. 65%   12. 5, -2, 3   13. 17, 3, 4, 14
                                        0, 2, 4   6, 12, 11, 9
                                        1, 6, -1   10, 8, 7, 13
                                                       5, 15, 16, 2

14. 1/5, 6   15. j, 4   16. $5 \frac{1}{4}$ inches

Problem Set 46

1. $5   2. -$1.54   3. -$1.25   4. even, a fair game   5. 23
6. 7   7. 15   8. 2   9. 6   10. 45% increase
11. 6% decrease   12. 100% increase   13. 1% decrease   14. 50% increase
15. 12   16. 6   17. 96   18. 66   19. 25%
20. 28   21. 8, 1, $1\frac{1}{2}$, $6\frac{1}{2}$
              $2\frac{1}{2}$, $5\frac{1}{2}$, 5, 4
              $4\frac{1}{2}$, $3\frac{1}{2}$, 3, 6
              2, 7, $7\frac{1}{2}$, $\frac{1}{2}$

Problem Set 47

1. 7   2. 35   3. 5   4. 48   5. $9.52   6. $2303.84
7. $2.50   8. $0.86   9. 1 5 10 10 5 1   10. 1/32   11. 31/32
12. $0.16   13. -$1   14. 40% increase   15. 7% increase
16. 8% decrease   17. 16% increase   18. 100% increase   19. 23%
20. 77%   21. 67   22. 20/87 = 23%   23. 190/3741
24. S(aturday)   25. D(ecember)

PROBABILITY
Answers to Problem Sets

Problem Set 48

1.
```
    1 1
   1 2 1
  1 3 3 1
 1 4 6 4 1
```

2. $0.44   3. -$1.25   4. $8

5. 30% increase   6. 43 pounds   7. 387 pounds

8. 94%   9. 6%   10. 42

11. 9, 2, $2\frac{1}{2}$, $7\frac{1}{2}$ 12. 3, -4, 1   13. 11 hours, 5 min.

$3\frac{1}{2}$, $6\frac{1}{2}$, 6, 5   -2, 0, 2

$5\frac{1}{2}$, $4\frac{1}{2}$, 4, 7   -1, 4, -3

3, 8, $8\frac{1}{2}$, $1\frac{1}{2}$

14. slightly more than eight days

Problem Set 49

1. 6.5   2. 5.2   3. 10   4. 50   5. -$1   6. 3/10

7. $8   8. 25% increase   9. 63   10. BLOB   11. $48

12. $432   13. $21.60   14. $6.40   15. 732   16. 95%   17. 5%

18. 733 pounds   19. 51%

Problem Set 50

1. 96   2. 5   3. 1 4 6 4 1   4. 1/16   5. 15/16   6. -$0.63

7. $25.20   8. $0.00, a fair game   9. 16   10. 30   11. 65%

12. 85%   13. 86%   14. 690   15. 50   16. 70   17. 83.5

18. 11%   19. 9, 2, $2\frac{1}{2}$, $7\frac{1}{2}$

$3\frac{1}{2}$, $6\frac{1}{2}$, 6, 5

$5\frac{1}{2}$, $4\frac{1}{2}$, 4, 7

3, 8, $8\frac{1}{2}$, $1\frac{1}{2}$

Problem Set 51

1. 7 minutes   2. 8 minutes   3. 70.9 cars   4. 70 cars   5. 6.4 cars

6. 7 cars   7. 19.3 cars   8. 7 cars   9. 62   10. 90

11. 240 pounds   12. 686 pounds   13. 926 pounds   14. 10 players

15. 92.6 pounds.   16. 1560 cards   17. 99.9%   18. 19%   19. 69%

20. 78 children

PROBABILITY
Answers to Problem Sets

Problem Set 52

1.

Brothers Sisters	Frequency
0	4
1	7
2	6
3	5
4	5
5	2
6	1

2. 23   3. 2   4. 17%   5. 43%

6. 2 students 7. 5 students 8. 13 students

9. 7.7%   10. 1.3 days   11. 1 day   12. 3.3

13. 3.6   14. 3   15. 3   16. 74

17. 8430 times

Problem Set 53

1. 82 per minute   2. 1.4 per second   3. $95.12   4. 27.3 feet

5. 2 pieces   6. 2 pieces   7. 5 pieces   8. 168 inches

9. 0 inches   10. 168 inches   11. 92   12. 3 students

13. 5 errors   14. 24 students   15. 2.4 errors   16. 3 errors

17. 5 errors   18. 33%   19. 30   20. 110

21. 110   22. 0   23. 2002

Problem Set 55

1. 5.5   2. 8.3   3. 7.5   4. 7.3   5. 6   6. 5.5

7. 8   8. 8.5   9. 7.5   10. 6   11. Gloomyos   12. Meely Oats

13. 4   14. 6   15. 8   16. 9   17. Scalia   18. 4

19. 4   20. 8   21. 2   22. Rehnquist   23. The Supreme Court

24. 363 is obtained   25. 55 is obtained   26. 93 people   27. 11

28. 259 pounds

Problem Set 56

1. $2,424,000   2. 3394 inches   3. 283 feet   4. 38 inches   5. 3 feet

6. 3443   7. 6 miles   8. 5:55   9.

6	0 1 2 4 9
5	1 1 1 2 4 5 5 6 6

10. 27 years

11. 55   12. 60.5   13. 10

14. 98   15. 60   16. 83.3   17. 86.5   18. 93   19. 6

20. 6   21. 14   22. 43%   23. 4   24. 7   25. 7.4

Problem Set 57

1.

11:00 – 1:00		4:00 – 6:00
	7	4 8
6 4 2 0	6	0 1 3 3 4 6 7 7 8
8 8 7 6 5 2 1 0 0	5	0 0 4 5 8 9 9
9 8 8 7 7 4	4	8 9
7	3	

2a. 29   b. 30   3a. 51.5   b. 60.5

4a. 58   b. 66.5   5. 31

6. Half weigh more than 175

7. The heaviest 25% weigh more than 200

8. 5:35   9. 6886   10. 1111 is obtained   11. 16 days   12. 6.5 days

13. 5.5 days   14. median   15. $2100   16. $7100   17. $887.50   18. 94%

PROBABILITY

Problem Set 58

1. 5 words/second
2. about 6 seconds
3. 200 pounds
4. 50 pounds
5. 175 pounds
6. 18 players
7. 9 players
8. 180 pounds
9. 25%
10. 18 players
11. 18 chapters
12. 10 pages
13. 29 pages
14. 22 pages
15. 28 pages
16. 15 pages
17. PICTURE
18. 7 mice
19. 7 mice
20. 5 mice
21. 5 students
22. 1 student
23. 3 tickets
24. 0 tickets
25. .5 tickets

Problem Set 59

1. 191 per minute
2. 3.2 per second
3. PICTURE
4. 22 years
5. 10 ghosts
6. 20 ghosts
7. $8
8. $4.50
9. $6
10. 50%
11. 75%
12. 30
13. 12 teams
14. 94
15a. 121.5
b. 126.5
16a. 128
b. 127
17a. 96
b. 116
18. PICTURE

Problem Set 60

1.
6	1
5	3
4	0 4
3	1 6 9
2	3 5 7
1	1 5 8 9

2. 50 grams
3. 29 grams
4. 40 grams
5. 19 grams
6. PICTURE
7. PICTURE
8. PICTURE
9. none
10. positive
11. positive
12. negative
13. none
14. positive
15. negative
16. $2,044,447
17. $4,770,376
18a. $318,025 yearly
b. $871 daily

Problem Set 61

1. 165 pounds
2.
18	3
17	4 6 7 7
16	1
15	2
14	1 6

3. 174 pounds
4. 177 pounds
5. 149 pounds
6. PICTURE
7. 4 hours
8. 6 students
9. 15 students
10. 1.9 hours
11. 2 hours
12. 11 students
13. 45
14. 10
15. 5 hours
16. yes
17. $3360
18. 3
19. $580
20. $567
21. $4225
22. $704
23. 800
24. 600
25. 75%
26. 50%

PROBABILITY
Answers to Quizzes

Quiz 1

1. 47% 2. 12% 3. 68% 4. 140% 5. 100% 6. 1
7. 12/25 8. 3.5 9. 100% 10. Answers vary.

Quiz 2

1.
Quarters	Dimes	Nickels
1	2	0
1	1	2
1	0	4
0	4	1
0	3	3
0	2	5
0	1	7
0	0	9

2. 3 3. 2 4. 44% 5. 111%
6. 4/5 7. 35% 8. 70%

Quiz 3

1. 94% 2. 80% 3. 18% 4. 2/15 5. 1/5 6. 4/5
7. 1 8. 1/5 9. 4/5 10. 1/5

Quiz 4

1. 1/4 2. 1/4 3. 1/8 4. 7/8 5. 1/2 6. 1/6
7. 5/6 8. 2/3 9. 1/46 10. 3/4

Quiz 5

1. 35% 2. 59% 3. 80% 4. 120% 5.
x	1	2	3	4
1	1	2	3	4
2	2	4	6	8
3	3	6	9	12
4	9	8	12	16

6. 1/4 7. 3/8 8. 5/16 9. 13/16

Quiz 6

1. 70% 2. 70% 3. 71% 4. 40% 5.
	2	3	4
2	2	3	4
3	3	3	4
4	4	4	4

6. 1/9
7. 5/9 8. 4/9 9. 8/9 10. 1

Quiz 7

1. 25% 2. ≈84% 3.
	2	3	4
2	7	7	7
3	7	8	8
4	7	8	9

4. 1/3 5. 5/9 6. 2/3
7. 5/9

Quiz 8

1.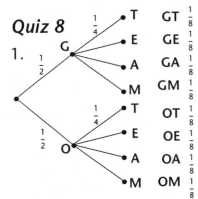

GT $\frac{1}{8}$
GE $\frac{1}{8}$
GA $\frac{1}{8}$
GM $\frac{1}{8}$
OT $\frac{1}{8}$
OE $\frac{1}{8}$
OA $\frac{1}{8}$
OM $\frac{1}{8}$

2. 1/8 3. 0 4. 1/2 5. 1
6. 1/4 7. 0

PROBABILITY
Answers to Quizzes

Quiz 9

1.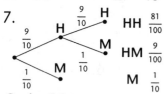

   AA $\frac{1}{16}$ AB $\frac{3}{16}$ BA $\frac{3}{16}$ BB $\frac{9}{16}$

2. B, B   3. 5/8   4a. 1   b. 2

Quiz 10

1. 78   2. 39   3. 234   4. 1   5. 2   6. 10

7. $8   8. 173

Quiz 11

1. 86   2. 43   3. 3   4. 6   5. 125   6. 10%

7. HH $\frac{81}{100}$ HM $\frac{9}{100}$ M $\frac{1}{10}$

Quiz 12

1. 20   2. 7   3. $6.40   4. $4.30   5. ≈4¢   6. $4.50

7. 165   8. 75%   9. 495

Quiz 13

1. 30%   2. 10%   3. 10%   4. 0%   5. 50   6. 70%

7. 30%

Quiz 14

1. 24, 29, 34   2. 38, 48, 47   3. 8, 3/8, 9   4. k, m, o   5. 73.4%

6. 28.4%   7. 39.0%   8. 74%   9. 11/46   10. 1/92

Quiz 15

1. 1, 6, 15, 20, 15, 6, 1   2. 1, 8, 28, 56, 70, 56, 28, 8, 1

3. 1, 9, 36, 84, 126, 126, 84, 36, 9, 1   4. 1/16   5. 1/16   6. 5/8

7. 5/16   8. 2 girls and 2 boys

Quiz 16

2. yes   3. $1.40   4. 40%   5. 60%

Quiz 17

1. $4   2. $60   3. $1   4.

11	1	15
13	9	5
3	17	7

PROBABILITY
Answers to Quizzes

Quiz 18

1.

Payoff	Cumulative
⁻$1	⁻$1
$2	$1
⁻$1	0
⁻$1	⁻$1
⁻$1	⁻$2

2. 15%   3. 28%   4. 6

Quiz 19

1. ⁻$.60   2. $31.15   3. ≈$1.33 per game   4. 14%

Quiz 20

1.

W/L	Profit	Cumulative
W	$11	$11
L	⁻$1	$10
L	⁻$1	$9

2. 76%   3. 75   4. 5%

Quiz 21

1. 7   2. 6   3. ≈7.1   4. 8   5. 10

Quiz 22

1. 10   2. 35   3. ≈29%   4. 6   5. 2   6. 5

7. ≈4.9

Quiz 23

1. 42   2. 32 years   3. 15   4. 32   5. $36 \frac{1}{2}$ 6. 10 min.

7. 74   8. Half of all NHL players are 26 or older and half are 26 or younger.

9. Answers vary.

Quiz 24

1. 100   2. 45   3. 80   4. impossible   5. impossible

6. 85   7. 50%   8. 25%   9. 0%   10. 50%

PROBABILITY
Answers to Tests

Test 1

1. 57%   2. 76%   3. 95%   4. 106%   5. 39%   6. 26%

7. less than 47%   8. more than 80%   9. more than 22%   10. less than 9%

11. more than 100%   12. more than 84%   13. 9/10   14. 3/4

15. 5/4   16. 13/100   17. 1   18. 30%   19. 9,997/10,000

20. about 97%   21. 2/11   22. 0   23. 1   24. 5/11   25. 4/11

26. 2/11   27. 7/11   28. 9/11   29. 25 times   30. 7   31. 28%

32. 25%   33. 3/4   34. 1/4   35. 1/4   36. 1/8   37. 7/8

38. 1/2   39. 1/2   40. 1/6   41. 1/3   42. 2/3   43. 47

44. 21   45. 14   46. 21/47   47. 14/47   48. 3/47   49. 33/47

50. 39/47   51. 0   52. 1   53. to 58.

59. 17   60. 4   61. yes   62. no

Dimes	Nickels	Pennies
1	1	2
1	0	7
0	3	2
0	2	7
0	1	12
0	0	17

Test 2

1.

+	1	2	3	4	5	6
1	2	3	4	5	6	7
2	3	4	5	6	7	8
3	4	5	6	7	8	9
4	5	6	7	8	9	10

2. 10   3. 5, 6, and 7   4. 9

5. 1/24   6. 1/2   7. 1/2   8. 1/4

9. 1/6   10. 7/12   11. 1   12. 5/24

13.

	1	2	3
1	9	9	9
2	9	10	10
3	9	10	11

14. 11   15. 9   16. 11

17. 50%   18. 94%   19. 69%   20. 80%   21. 8%

22. 86%   23. 20%   24. 137%   25. 112%   26. 50%

27. ≈77%   28. 7   29. ≈23%   30. 74%   31. $64   32. ≈77%

33. 0%   34. 35   35. 32   36. HHHHH or TTTTT   37. 3¢

38. 3   39.   40. 1/8   41. 0   42. 1/4

43. 1/4   44.   45. 1/9

46. 1/3   47. 1/3

48. 5/9   49. 1

50. 1/3

Test 3

1.

AA $\frac{9}{16}$ ≈ 56%
AB $\frac{3}{16}$ ≈ 19%
BA $\frac{3}{16}$ ≈ 19%
BB $\frac{1}{16}$ ≈ 6%

2. A, A
3. 5/8
5. G, G
6. 5/9
7. 11

4.

RR $\frac{1}{9}$ ≈ 11%
RG $\frac{2}{9}$ ≈ 22%
GR $\frac{2}{9}$ ≈ 22%
GG $\frac{4}{9}$ ≈ 44%

8. ≈6   9. ≈78   10. ≈39

11.

RR	$\frac{1}{10}$
RG	$\frac{3}{10}$
GR	$\frac{3}{10}$
GG	$\frac{3}{10}$

12. 3/5　　13. 2　　14. 1　　15. 7

16.

H	$\frac{2}{3}$
MH	$\frac{2}{9}$
MMH	$\frac{2}{27}$
MMM	$\frac{1}{27}$

17. ≈3.7%　　18. ≈96.3%

19. 63%　　20. 44%

21. 30%

22. problem 21

23. ≈5 times　　24. ≈20 times　　25. ≈1 time　　26. once

27. 10　　28. 2　　29. ≈67%　　30. ≈33%　　31. 26　　32. 54

33. 270　　34. 135　　35. 540　　36. 4　　37. 36　　38. 90%

39. $4　　40. $284.10　　41. $6

42. E = 1, K = 0; R = 2, 3, 4, 6, 7, or 8; T = 8, 7, 6, 4, 3, or 2

43. 1089
　　x 9
　　9801

Test 4

1.　11, 13　　2.　21, 25　　3.　96, 192　　4.　d, v　　5.　3/4, 10/13

6.　86.3%　　7.　12.4%　　8.　5.6%　　9.　17.0%　　10. 15　　11. 8

12. HEDGE　　13. 1　　14. 0　　15. 1, 5, 10, 10, 5, 1　　16. 1/64

17. 15/64　　18. 15/64　　19. 7/64　　20. 5/16　　21. 88%　　22. 12

23. 6　　24. 84　　25. 102　　26. ≈46　　27. 1/4　　28. 1/16

29. 1/64　　30. 3/64　　31. 1/32　　32. 3/14　　33. 3/56　　34. 0

35. 3/56　　36. 1/28　　37. 300　　38. 60　　39. 210　　40.25%

41. ≈96.3%　　42. $1.10　　43. $3.30　　44. 25%　　45. 12%　　46. 6%

47. 96%　　48. ≈93.6%　　49. $12　　50. $36

Test 5

1.　130.9 pounds　　2.　$3.50　　3.　$78.75　　4.　$3 per game

5.　$.80 per game　　6.　1/4　　7.　3/4　　8.　$0 (a fair game)

9.　1/9　　10. 8/9　　11. ≈$.22　　12. 1 4 6 4 1　　13. 3/8　　14. ⁻$2.50

15. 4999/5000　　16. 60　　17. 540　　18. 90%　　19. 50

20. 86%　　21. 472　　22. ≈16.7%　　23. ≈83.3%　　24. 812　　25. SLIDE

26. 12%　　27. 100%　　28. 25%　　29. ≈4.6%　　30. 8 pounds　　31. 10 pounds

32. 17 pounds　33. 154 inches　　34. 4　　35. 2　　36. 2.6 or 3

37. 13, ⁻15　　38. a, S

39.

4	$\frac{1}{2}$	3
$1\frac{1}{2}$	$2\frac{1}{2}$	$3\frac{1}{2}$
2	$4\frac{1}{2}$	1

40.

13	⁻1	0	10
2	8	7	5
6	4	3	9
1	11	12	⁻2

PROBABILITY
Answers to Tests

Test 6

1. 4515.5 miles   2. 244 miles   3. 15 miles   4. 13   5. 12.1
6. 14   7. mean   8. 66   9. 6   10. 86   11. 12.3
12. 13   13. 47   14. 30 hours   15. 32 hours   16. 43 hours   17. 19 hours
18. 19   19. 11   20. 12   21. 25%   22. 7 days   23. 14 days
24. 5   25. 4   26. 4   27. 17   28. 1   29. F
30. T   31. T   32. T   33. T   34. F   35. T
36. T   37. F   38. F   39. T   40. 30   41. 15
42. 20   43. yes   44. 20 miles   45. 6:56   46. 363
47. Answers vary.